# PRACTICE TESTS GCSE BIOLOGY

June Hassall PhD

Nelson

Thomas Nelson and Sons Ltd
Nelson House    Mayfield Road
Walton-on-Thames    Surrey
KT12 5PL    UK

51 York Place
Edinburgh
EH1 3JD    UK

Thomas Nelson (Hong Kong) Ltd
Toppan Building 10/F
22A Westlands Road
Quarry Bay    Hong Kong

Distributed in Australia by

Thomas Nelson Australia
480 La Trobe Street
Melbourne    Victoria 3000
and in Sydney, Brisbane, Adelaide and Perth

© J. Hassall, 1988

First published by Thomas Nelson and Sons Ltd 1988

ISBN 0–17–448157–8

NPN 987654321

Printed and bound in Great Britain by
Butler & Tanner Ltd, Frome and London

All Rights Reserved. This publication is protected in the United
Kingdom by the Copyright Act 1956 and in other countries by comparable
legislation. No part of it may be reproduced or recorded by any means
without the permission of the publisher. This prohibition extends
(with certain very limited exceptions) to photocopying and similar
processes, and written permission to make a copy or copies must
therefore be obtained from the publisher in advance. It is advisable
to consult the publisher if there is any doubt regarding the legality
of any proposed copying.

# PREFACE

This book is designed to help students to revise for the new GCSE examinations in biology. All the syllabuses of the examination boards and groups in England, Scotland, Wales and Northern Ireland have been used in identifying topics to be tested.

The questions sample the full content area of the syllabuses and most of the practical skills listed for teacher assessment. There is a particular stress on topics most often tested in examinations, on difficult topics and on topics and skills which are new to the syllabuses.

The questions are arranged in sixteen tests, and in each test there are multiple choice questions and longer questions. Answers to the multiple choice questions and full marking schemes for the longer questions are included.

In the preparation of this book I have been helped by several people. I would like to thank my daughter, Catherine Mitchelmore, for commenting on the questions and assisting with the item analysis, and also John Finagin who commented on particular questions, marking schemes and support materials. I am very grateful to the following teachers who gave the tests to their students; they gave me useful feedback which has helped to improve the tests:

Mr A. Davies, H.O.S. Oldfield Girls School, Bath;
Mr A. F. C. H. Doig, St Andrews College, Glasgow;
Mr J. Finagin, Rhyl High School, Rhyl, Wales;
Dr M. Hudson, Oulder Hill Community School, Rochdale, Lancs;
Mrs B. Parkin, Fernwood School, Wollaton, Nottingham;
Mrs L. Coppin, Bexley/Erith Technical High School, Kent.

I trust this book will prove to be a useful source of questions and answers, and I would be interested to receive, via Nelson and Sons, any comments and class test results from teachers.

Finally, it is a pleasure to thank Mr Chris Coyer and Ms Sharon Jacobs for their work in the production and execution of this project. Special thanks are given to Sharon for her assistance in the final stages.

June Hassall
August 1987

# CONTENTS

How to use this book     1

Descriptions of process and practical skills     2

| Questions | | Answers | |
|---|---|---|---|
| Test 1 | 10 | Test 1 | 110 |
| Test 2 | 17 | Test 2 | 111 |
| Test 3 | 23 | Test 3 | 112 |
| Test 4 | 29 | Test 4 | 114 |
| Test 5 | 35 | Test 5 | 116 |
| Test 6 | 40 | Test 6 | 117 |
| Test 7 | 47 | Test 7 | 119 |
| Test 8 | 52 | Test 8 | 120 |
| Test 9 | 60 | Test 9 | 122 |
| Test 10 | 66 | Test 10 | 124 |
| Test 11 | 72 | Test 11 | 126 |
| Test 12 | 77 | Test 12 | 128 |
| Test 13 | 83 | Test 13 | 130 |
| Test 14 | 89 | Test 14 | 133 |
| Test 15 | 95 | Test 15 | 135 |
| Test 16 | 102 | Test 16 | 137 |

Appendix I    Content areas covered, and levels of difficulty of the questions     140

Appendix II    Process and practical skills covered, and levels of difficulty of the questions     144

Appendix III    Teacher use of test questions     148

# HOW TO USE THIS BOOK

## Students

* Use some of the questions to test yourself on particular **topics** as you study and revise your biology course. Use Appendix I on pp. 140–3 to choose the topics and pick out easy, average or hard questions to test yourself, and then mark your answers.
* Use some of the questions to help you with process and practical **skills** which may be new to you. The skills and how they may be tested are described on pp. 2–8. Use Appendix II on pp. 144–7 to choose the skills on which you want to be tested. Answer the questions and then mark them.
* You can also use an entire test for **revision** because each test samples the whole course. Do each test in turn. The first eight tests are easier. Try these first, and mark each one as you go along. Try and learn from your mistakes. Then you can try the harder tests 9–16. With the marking schemes for the tests you will be given an idea of the approximate grade range you may get on your final examination.

## Teachers

* The tests as presented can be used as **revision tests**. They are of two main kinds: tests 1–8 for students expecting to achieve grades C–G, and tests 9–16 for students expecting to achieve grades A–E. In the marking schemes there is an idea of the approximate grade range on GCSE which corresponds with each score.
* The book can also be used as a **bank of questions**. Each question is described as to its content area, the skill tested and difficulty level. These descriptions are in Appendices I and II on pp. 140–7. These appendices can be used to identify questions with particular characteristics which can then be included in **class tests** or end-of-term tests. Some guidelines for designing tests are given in Appendix III on pp. 148–56.
* Appendix III also describes how to use class test results to work out the **difficulties** and **discriminations** of test questions.

# DESCRIPTIONS OF PROCESS AND PRACTICAL SKILLS

The various examination boards and groups list the skills to be tested in the teacher assessment part of the examination. For comparison the main headings of these are given below:

**London and East Anglian Group (LEAG)**
A. Make and record accurate observations.
B. Perform experiments and interpret the results.
C. Design and evaluate an experiment.

**Midland Examining Group (MEG)**
1. Following instructions.
2. Handling apparatus and materials.
3. Observing and measuring.
4. Recording and communicating.
5. Interpreting data.
6. Experimental design/ problem solving.

**Northern Examining Association (NEA)**
1. Measurement.
2. Observation.
3. Handling materials and apparatus.
4. Recording.
5. Data and its interpretation.
6. Design.

**Northern Ireland Schools Examination Council (NISEC)**
I   Manipulative skills.
II  Following instructions.
III Observation and measurement.
IV  Recording and presenting information.
V   Experimental design/ problem solving.

**Scottish Examining Board (SEB)**
Designing an investigation.
Carrying out experimental procedures.
Recording results.
Analysing results and drawing conclusions.
Simple experimental techniques.

**Southern Examining Group (SEG)**
1. To follow written and diagrammatic instructions.
2. To handle apparatus and materials.
3. To make and convey accurate observations.
4. To record results in an orderly manner.
5. To formulate an hypothesis.
6. To design an experiment to test an hypothesis.
7. To carry out safe working procedures.

**Welsh Joint Education Committee (WJEC)**
1. Observational and recording skills.
2. Measurement skills.
3. Procedural skills.
4. Manipulative skills.
5. Formulation of hypotheses and experimental design.

Some of these practical skills can also be partly tested on the written papers. In addition all questions will also involve some amount of recall and understanding, and some may test application to new situations.

The following list of process and practical skills has been developed from the analysis of the various syllabuses and specimen questions. Each of the questions in this book has been designed to mainly cover one of these skills areas (these lists can be found in Appendix II).

# Recall and understanding

This is the basic acquisition of knowledge. It covers remembering (recalling) and grasping the meaning of (understanding) basic facts, definitions, ideas and principles of the whole course. You have to show that you understand what you have been taught.

# Application

This is the use of or transfer of information to a new situation. You may have learned information in a certain way, but you cannot just regurgitate it. You will have to reorganize it in a new way, or use information to explain results which are new to you. It covers the application of what you have learned in school to everyday life, industry, technology, etc.

## Obtaining and interpreting information

Information can be given to you in many different forms. You have to explain it (translate or interpret it) to show that you fully understand it. For example you will have to deal with these different kinds of information:
(a) Written material related to the course – identify the main themes and important ideas and explain the information.
(b) Tables – select particular readings and identify patterns in the results.
(c) Diagrams – give the names and functions of labelled parts, and describe their importance/significance.
(d) Flow charts – name the processes labelled and their importance, and show that you understand what the flow chart is about.
(e) Line graphs – select particular readings, identify patterns in the results and suggest other values (make predictions).
(f) Bar charts/graphs – select particular readings, e.g. heights of the bars.
(g) Pie charts – show that you understand the significance of the various sectors shown on the pie charts.

## Presenting information and recording results

You will either be given information, have recorded it from an investigation or will have had to remember it. You will then have to present it in one of the following forms:
(a) Written material – you will have to answer short questions or be asked to write a longer account about a certain topic. You may be given the sub-headings to use, or you may have to decide upon your own.
(b) Tables – you will either have been given written information or diagrams, or figures (data/readings) or have collected your own. You will then have to enter them in a table you have been given or have had to make for yourself. The correct display of information is important.
(c) Diagrams – from information given, or from memory.
(d) Flow charts – from information given, e.g. preparing a food web.
(e) Line graphs – from information given or collected from an investigation. You may be given the axes and/or the scales of the graph or have to work them out for yourself.
(f) Bar charts/graphs – as for line graphs.
(g) Pie charts – from information given you will have to work out the sectors to represent each of the parts of the pie chart.

# Dealing with numbers

This can be a part of obtaining and interpreting information, or of presenting information and recording results. However, dealing with numbers is a skill which students often find difficult and so it has been treated separately. It tests 'number sense'.

You will usually be given information (data/readings) and have to deal with it in the following ways:
(a) Arithmetic operations of +, −, × and ÷, for example using figures from a table, or deciding upon the magnification of a drawing.
(b) Calculation of average (mean) of a set of numbers.
(c) Calculations of percentages, and conversion of percentages to whole numbers.
(d) Calculations of proportions for comparison of data, and expected proportions of different offspring in genetic crosses.
(e) Calculations of ratios for comparison of data, and expected ratios of different offspring in genetic crosses.
(f) Calculations of growth rates from a graph.
(g) Determination of density and percentage cover from sampling of organisms.
(h) Finding the significance of or pattern in a set of numbers.

# Experimental procedure

This is the actual carrying out of investigations. It covers many sub-skills, many of which can only be assessed by your teacher during the course of practical work. However, you may have to describe on the written paper that you know how to do them and so they are listed here.

## (a) Observation

Observation involves using the senses to describe organisms, and identify changes, similarities and differences, etc. The observation is the major skill, and recording it is of secondary importance. Some examples are:
1. Observing the whole or part of an organism with the use of a hand lens or the microscope.
2. Observing changes, e.g. in the colour of test solutions in food tests, and production of gas bubbles during photosynthesis.
3. Finding similarities between organisms (matching). This skill is also tested when identifying organisms using a biological key when you have to observe the organisms and match them to the descriptions in order to classify them.

4. Finding differences between organisms. As for similarities, you may be given actual specimens, diagrams, photographs, etc.

## (b) Measuring

This involves the correct use of measuring instruments, the reading of the scales and the recording of the results together with the appropriate units. In the syllabuses the degree of accuracy required will be listed, for example:
1. Thermometer. Allow 3 minutes for equilibration (coming to the same temperature as the surroundings). Read to the nearest 0.5 or 1°C.
2. Ruler or metre stick. Read to the nearest 1 mm (0.1 cm).
3. Measuring cylinder. Place it on a flat surface, and, with the eye level with the water level, read the lowest part of the curve (the meniscus). Read to the nearest 1 $cm^3$.
4. Stop-clock. Adjust it to the zero mark before starting. Read to the nearest second (s).
5. Lever-arm balance. Mass. Read to the nearest 1 gram (g).

## (c) Making drawings

Drawing in this sense is taken to mean the accurate representation of a biological specimen, with correct proportions and to a certain scale. It is contrasted to making a diagram, which is just a useful sketch which may not look very much like 'the real thing'. Some examples are the drawing of the actual hind leg of an insect, and drawing the pattern of the main veins of a leaf.

## (d) Procedural skills

You should be able to follow instructions that have been spoken, written or given as a diagram (i.e. flow chart). This will include field work, microscope work and general laboratory work. You should also be able to arrange the steps in an activity or from written information in the correct order (sequence) so that they make sense. You may also have to describe or account for the steps in a certain procedure.

## (e) Handling materials and apparatus correctly and safely

Some examples of what you may have to do are: handle and assemble apparatus, handle and use the microscope, show respect and caution when handling living organisms, correctly use biological instruments and chemicals and carry out procedures such as heating and filtering. You may also have to give an account of how you would do these things.

# Designing and evaluating

This skill depends mainly upon the correct setting up of investigations and proper use of controls, so that when you get some results you can be fairly sure that they mean something and are not just due to chance!

## (a) Designing (planning) investigations

You have to decide upon or will be given an hypothesis (an untested theory) which may or may not be correct. You have to plan an investigation to test it. These points should be borne in mind:
1. Select or list the necessary apparatus, materials and chemicals.
2. Describe how they might be used in the investigation.
3. Carefully describe how the control should be set up so as to differ from the experiment only in the one factor under investigation for, e.g.:
   (a) Use large numbers (e.g. of seeds or fruits) so that any differences in the results are not likely to be due to chance.
   (b) Use the same masses or volumes for proper comparison.
   (c) Set up under the same conditions, e.g. of light, temperature, etc.
   (d) Take the same number of readings and at the same times.
4. Describe the main steps in the procedure and any special techniques.
5. Decide ahead of time how you will judge a good result, e.g. when seeds have germinated, when a reaction has been completed, etc.
6. Mention that the investigation should be repeated with other samples.

## (b) Criticizing (evaluating) investigations

To criticize is to 'judge the value of', whether negative or positive. You may have to criticize or evaluate an investigation you have planned yourself, or one which has been described to you. For example you may need to:
1. Account for the use of a certain piece of apparatus or chemical or a certain procedure in an investigation.
2. Criticize a diagram of the apparatus used in an investigation.
3. Describe problems which might be expected.
4. Identify factors which might affect the 'fairness' of the test, and say how they could be minimized (see also controls above).
5. Identify possible sources of error (inaccuracies) and say how they could be corrected.

## (c) Drawing conclusions and making generalizations

You may have to examine the results of your own investigation or you may be given information/data in written form, or as a table, diagram, line graph,

bar chart, etc. For example you may have to:
1. Give reasons for the results.
2. Say what was the aim of the investigation.
3. Say what conclusions can properly be drawn from the results. (This will again depend upon there having been proper controls.)
4. Comment on the limitations of the conclusions that can be drawn.
5. Use conclusions to suggest what might happen in closely related situations.
6. Describe other experiments which might be necessary before a conclusion can be drawn.
7. Generalize (form a conclusion which has wider application or usefulness).

# QUESTIONS

The questions are arranged in the form of sixteen tests.

You can: *either* choose questions of a certain type from several tests (by referring to Appendix I (for particular content or subject matter) or Appendix II (for particular skills)),
  *or* do an entire test as a revision of the course. (Tests 1–8 are easier than Tests 9–16.)

## On each test

*Questions 1–16*
1. These are multiple choice questions. They have five suggested answers *(A), (B), (C), (D)* or *(E)*. Read each question *carefully* and decide which of the possible answers is the best one.
2. On an answer sheet write down the number of the test and the numbers 1–16. Write your choice, *(A), (B), (C), (D)* or *(E)*, next to the corresponding number on your piece of paper.
3. If you are choosing questions from several tests, write down the full question number, e.g. 3:12 (Test 3 question 12), so that you will have the reference for later on when you come to look at the answers.

*Questions 17–20*
1. These are longer questions. Each question is divided into several parts, and the number of marks for each part is shown.
2. Again, if you are answering questions from several tests, write down the full question number.
3. If the question involves the making of a bar chart or line graph, the graph paper you will need has been included in the book for you to use.
4. If the question involves writing out an answer, you should use a separate sheet of writing paper.
5. As you answer each part of the question make sure that you write down (a), (b), (c), etc., to show which part you are answering.
6. For some questions you may need to use a ruler, and you may use a calculator where appropriate (but make sure to write down on your answer sheet the steps in the calculation you have performed).

# Test 1

*(Answers on p. 110)*

**1:1** Which group of vertebrates has four limbs, is cold blooded and has scales?

(A) Fish
(B) Amphibia
(C) Reptiles
(D) Birds
(E) Mammals

**1:2** Which of the following processes does **not** release moisture into the air?

(A) Sweating
(B) Photosynthesis
(C) Transpiration
(D) Breathing
(E) Evaporation

**1:3** In mammals gaseous exchange occurs in the:

(A) larynx.
(B) trachea.
(C) bronchi.
(D) alveoli.
(E) diaphragm.

**1:4** Which of the following would lead to glucose being excreted in the urine?

(A) Eating a large meal with extra protein
(B) Eating a small bar of chocolate
(C) Eating a large bar of chocolate
(D) Having too little insulin produced
(E) Having too much insulin produced

**1:5** Which of the following pairs of items is **not** correctly matched?

(A) Implantation – embedding of fertilized egg
(B) Fertilization – joining of egg and sperm
(C) Ovulation – release of semen from the penis
(D) Semen – fluid containing sperm
(E) Sperm – male gamete

**1:6** Which group of plants has a simple structure with no proper stems, roots and leaves?

(A) Algae
(B) Ferns
(C) Conifers
(D) Monocotyledons
(E) Dicotyledons

**1:7** Which of the following represents a complete and correct food chain?

(A) Pondweed→tadpoles→small fish→large fish
(B) Plankton→slugs→small birds→large fish
(C) Grass→grasshoppers→cows→humans
(D) Cow→human
(E) Grass→fish→human

**1:8** The conditions found in a certain place are called the:

(A) habitat.
(B) population.
(C) community.
(D) environment.
(E) ecosystem.

**1:9** In the carbon cycle, which of the following processes adds carbon dioxide to the atmosphere?

I  Decay
II  Fermentation
III  Combustion
IV  Photosynthesis

(A) I and II only
(B) I, II and III only
(C) II, III and IV only
(D) II and III only
(E) I, II, III and IV

**1:10** The table below shows the time taken for the breakdown of starch with an enzyme at different pH levels.

| pH | 5 | 5.5 | 6 | 6.5 | 7 | 7.5 | 8 |
|---|---|---|---|---|---|---|---|
| Time taken in minutes | 6 | 4 | 3 | 2 | 1.25 | 1.25 | 3 |

What is the best pH range for this enzyme?

(A) pH 5–5.5
(B) pH 5.5–6
(C) pH 6–6.5
(D) pH 6.5–7
(E) pH 7–7.5

**1:11** Which of the following arteries carries deoxygenated blood?

(A) Renal artery
(B) Hepatic artery
(C) Aorta
(D) Pulmonary artery
(E) Coronary artery

**1:12** Five test-tubes were set up as shown below, and placed in bright sunlight for an hour. In which test-tube would most carbon dioxide be given out?

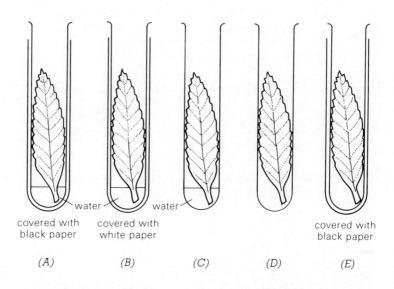

**1:13** Anaerobic respiration is used in the production of which of the following?

    I    Bread
    II   Vinegar
    III  Wine
    IV  Methane (biogas)

(A) I and II only  
(B) I and III only  
(C) I, II and III only  
(D) I, III and IV only  
(E) I, II, III and IV  

**1:14** Which of the following statements about growth are correct?

    I    In growing animals cell division usually takes place all over the body.
    II   In growing plants cell division takes place mainly in shoot and root tips.
    III  Mature organisms grow at the same rate as immature ones.
    IV  Growth is a characteristic of all living things.

(A) I, II and III only  
(B) II, III and IV only  
(C) I, II and IV only  
(D) I, III and IV only  
(E) I, II, III and IV

**1:15** Which of the following things happen in order to **decrease** heat loss from the skin?

(A) More sweat evaporates.
(B) Blood capillaries in the skin dilate.
(C) The hairs stand upright.
(D) More blood is brought to the skin.
(E) The liver releases more heat.

**1:16** Human body cells contain 46 chromosomes. How many chromosomes are present in an unfertilized human egg cell inside the female?

(A) 22 pairs of chromosomes plus X and Y
(B) 22 pairs of chromosomes plus two X chromosomes
(C) 22 pairs of chromosomes plus one X chromosome
(D) 22 chromosomes plus an X or Y chromosome
(E) 22 chromosomes plus an X chromosome

**1:17** The table below shows some of the ingredients of an egg, bacon and mushroom flan.

| Ingredients | Typical amounts | |
| --- | --- | --- |
| | Per 100 g | Per half-flan serving |
| Energy | 250 kcal<br>1045 kJ | 500 kcal<br>2090 kJ |
| Protein | 9.5 g | |
| Carbohydrate | | 38 g |
| Total fat | 15.6 g | |
| Unsaturated fat | | 4.6 g |
| Added salt | | 1.6 g |

(a) What is the mass of:
   (i) a half-flan serving? *(1)*
   (ii) the protein in a half-flan serving? *(1)*
   (iii) the total fat in a half-flan serving? *(1)*

(b) What is the mass of:
   (i) carbohydrate in 100 g of flan? *(1)*
   (ii) unsaturated fat in 100 g of flan? *(1)*
   (iii) salt in 100 g of flan? *(1)*

(c) (i) What is the percentage of protein in 100g of flan? *(1)*
   (ii) What is the percentage of protein in a half-flan serving? *(1)*

**1:18** Read the following paragraphs and then:
(a) fill in the table, and
(b) answer the questions below.

'The burning of fossil fuels causes many problems. Sulphur dioxide and oxides of nitrogen are produced which form part of 'acid rain'. In addition the incomplete combustion of the fuels causes the release of carbon (soot) into the air. Using smokeless fuels would help to prevent these problems, and special extractors can be built in factories to remove sulphur dioxide from the air.

Another problem arises when the nitrates from excess fertilizers are washed from the land into fresh water. In addition to their effect on life in water, excess nitrates can also be harmful to babies.'

(a) Use the information above to fill in a table like that below. *(4)*

| Pollutant | Source | Control |
|-----------|--------|---------|
|           |        |         |
|           |        |         |
|           |        |         |
|           |        |         |

(b) Using the information answer these questions:
  (i) Name two pollutants which form part of 'acid rain'. *(2)*
  (ii) Name a pollutant which could be a health hazard to humans. *(1)*
  (iii) Name a pollutant which it would be expensive to prevent from entering the air. *(1)*

**1:19** (a) Use the key provided to identify each of the five arthropods shown below.

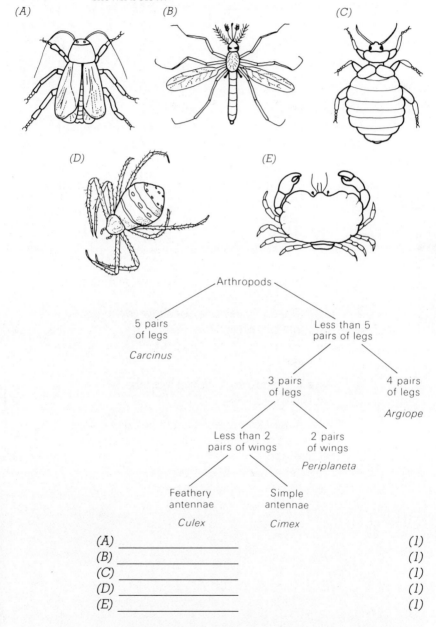

| (A) | _____ | (1) |
| (B) | _____ | (1) |
| (C) | _____ | (1) |
| (D) | _____ | (1) |
| (E) | _____ | (1) |

(b) (i) Using the key write down a description of *Culex* sp. *(3)*
    (ii) Which of these features is characteristic of insects? *(1)*

**1:20** The following experiment was set up and the results recorded.

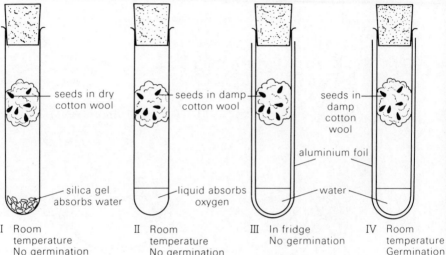

I  Room temperature
   No germination

II Room temperature
   No germination

III In fridge
    No germination

IV Room temperature
   Germination

(a) (i) What conclusion can be drawn by comparing test-tubes I and IV? *(2)*

(ii) What conclusion can be drawn by comparing test-tubes II and IV? *(2)*

(iii) What conclusion can be drawn by comparing test-tubes III and IV? *(2)*

(b) Why is it necessary to set up test-tube IV? Explain your answer. *(2)*

(c) What other experiments should be done before we can be sure of the conditions necessary for germination? *(1)*

# Test 2

*(Answers on p. 111)*

**2:1** Which group of vertebrates has scales, is cold blooded (poikilothermic) and lives on land?

(A) Fish
(B) Amphibia
(C) Reptiles
(D) Birds
(E) Mammals

**2:2** Look at the food chain shown below.

X →tadpole →water beetle →fish →man

X could be:

(A) a frog.
(B) a tree.
(C) grass.
(D) pondweed.
(E) seaweed.

**2:3** Which of these structures is found in plant cells but **not** in animal cells?

(A) Cell membrane
(B) Cell wall
(C) Cytoplasm
(D) Nucleus
(E) Vacuole

**2:4** What is the function of the white blood cells called phagocytes? They:

(A) carry oxygen around the body.
(B) carry carbon dioxide around the body.
(C) engulf invading micro-organisms.
(D) heal cuts and grazes.
(E) produce antigens against micro-organisms.

**2:5** Look at the diagram below of a section of a structure in the human body.

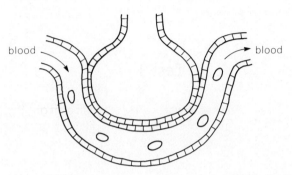

Where in the body would you expect to find this structure? In the:

(A) skin.
(B) lungs.
(C) kidneys.
(D) stomach wall.
(E) heart.

**2:6** Examination of which of the following could **not** be used to distinguish between monocotyledons and dicotyledons?

(A) Cotyledons
(B) Roots
(C) Petals
(D) Leaf shape
(E) Leaf venation

**2:7** In the carbon cycle, which of the following processes removes carbon dioxide from the atmosphere?

(A) Decay
(B) Fermentation
(C) Respiration
(D) Photosynthesis
(E) Combustion

**2:8** A stream used to contain algae and fish. A factory was set up near to the stream and the algae increased enormously in numbers. After a while the fish began to die. The most likely reasons for this are:

I  The algae choked the gills of the fish.
II  Poisonous chemicals were emptied into the stream.
III  Decay caused a lack of oxygen for the fish.
IV  The fish ran out of food.

(A) I and II only
(B) I and III only
(C) II and III only
(D) I and IV only
(E) II and IV only

**2:9** Which of the following diseases is caused by a pathogenic organism?

(A) Sickle cell anaemia
(B) Malaria
(C) Diabetes
(D) Haemophilia
(E) High blood pressure

**2:10** Which of the following is the correct definition of osmosis?

(A) The movement of water against a diffusion gradient
(B) The movement of solutes along a diffusion gradient
(C) The movement of particles from a place of low concentration to a place of high concentration
(D) The movement of solutes, through a (selectively) semi-permeable membrane from a weak solution to a strong solution
(E) The movement of a solvent, from a weak solution to a strong solution through a (selectively) semi-permeable membrane

**2:11** Which part of the body makes the juice which helps to neutralize the acid contents coming from the stomach?

(A) Liver
(B) Gall bladder
(C) Stomach
(D) Salivary glands
(E) Small intestine

**2:12** Which of the following pathways for the circulation of blood in a mammal is correct?

(A) Left ventricle→body→right atrium (auricle)
(B) Left ventricle→lungs→left ventricle
(C) Right atrium (auricle)→lungs→left atrium (auricle)
(D) Right atrium (auricle)→body→left ventricle
(E) Right atrium (auricle)→lungs→left ventricle

**2:13** Which of the following statements about the eye are **correct**?

I   The cornea bends light rays.
II  The choroid contains many blood vessels.
III The ciliary muscles change the pupil diameter.
IV  The cones respond to bright light.

(A) I, II and III only
(B) I, II and IV only
(C) II and IV only
(D) II, III and IV only
(E) I, II, III and IV

**2:14** Which of the following is **not** essential for germination?

(A) Suitable temperature
(B) Oxygen
(C) Water
(D) Food store
(E) Light

**2:15** A woman has regular menstrual cycles of 28 days. It is assumed that she ovulates on the 15th day of her cycle.

If sperm can live in the female for 3 days, and an egg can live for 2 days, on which days of the cycle could sexual intercourse result in pregnancy?

(A) Days 12–16 only
(B) Days 12–17 only
(C) Days 13–16 only
(D) Days 15–16 only
(E) Day 15 only

**2:16** Which of the following pairs of items is **not** correctly matched?

(A) Genotype — alleles which an organism contains
(B) Phenotype — external expression of alleles
(C) Homozygous — identical alleles for a particular characteristic
(D) Heterozygous — different alleles for a particular characteristic
(E) Incomplete dominance — one allele is dominant to the other

**2:17** The diagram opposite shows a microscope.

(a) A student is going to examine a transverse section of a stem under the low power of the microscope. Give the correct sequence for doing this from the steps listed below. *(3)*

    (i) Looking from the side use the coarse adjustment to lower the body tube close to the slide.
    (ii) Put in the eyepiece and objective lenses.
    (iii) Adjust the condenser so its top is level with the top of the stage.
    (iv) Adjust the mirror so that it collects light.
    (v) Put the slide on the stage and attach it with clips.
    (vi) Use the fine adjustment to raise the body tube to focus the object.

(b)　List the names of the parts (A)–(H)　　(4)

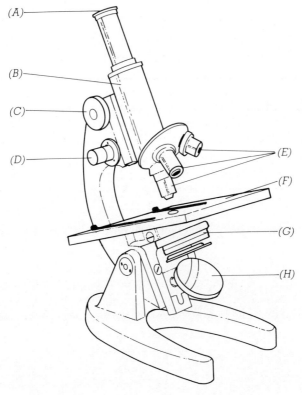

(c)　(i)　What is the correct way to carry a microscope?　(1)
　　(ii)　How should the lenses be cleaned?　(1)

**2:18**　(a)　(i)　Define continuous variation.　(1)
　　　　　(ii)　Give an example.　(1)
　　　(b)　(i)　Define discontinuous variation.　(1)
　　　　　(ii)　Give an example.　(1)
　　　(c)　(i)　What name is given to a change in the genes/alleles?　(1)
　　　　　(ii)　Give an example.　(1)
　　　(d)　(i)　What causes Down's syndrome?　(1)
　　　　　(ii)　Down's syndrome babies are more common if the mother is over 40 years. What could account for this?　(1)

**2:19** (a) 120 g of a certain breakfast cereal contained 30 g of fibre.

    (i) What is the percentage of fibre in this cereal? *(2)*

    (ii) How much fibre would there be in an average serving of 40 g of this cereal? *(2)*

(b) The table below shows the composition of 100 cm³ of full-cream milk and 100 cm³ of semi-skimmed milk.

|  | *Full-cream milk* | *Semi-skimmed milk* |
| --- | --- | --- |
| Energy | 280 kJ | 200 kJ |
| Protein | 3.4 g | 3.5 g |
| Carbohydrate | 4.8 g | 5.0 g |
| Total fats | 3.9 g | 1.6 g |
| Saturated fats | 2.5 g | 1.0 g |

    (i) How much more total fat would there be in a litre of full-cream milk compared with a litre of semi-skimmed milk? *(2)*

    (ii) A pint is 568 cm³. How much carbohydrate would there be in a pint of semi-skimmed milk? *(3)*

**2:20** The diagram below shows how water is recycled in nature.

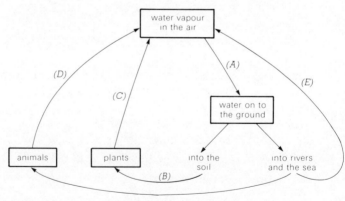

(a) Name the processes labelled *(A)*, *(B)*, *(C)*, *(D)* and *(E)*. *(5)*

(b) What is the importance of process *(C)* to plants? *(1)*

(c) Give the letters of all the processes which could involve evaporation. *(2)*

# Test 3

*(Answers on p. 112)*

**3:1** Decay is brought about by:

- (A) producers.
- (B) consumers.
- (C) scavengers.
- (D) decomposers.
- (E) herbivores.

**3:2** Respiration produces:

- I   energy.
- II  carbon dioxide.
- III oxygen.
- IV  excretory products.

- (A) I, II and III only
- (B) II, III and IV only
- (C) I, II and IV only
- (D) I, III and IV only
- (E) I, II, III and IV

**3:3** The pancreas produces:

- (A) adrenaline.
- (B) insulin.
- (C) thyroxine.
- (D) progesterone.
- (E) testosterone.

**3:4** Which of the following structures is found in both males and females?

- (A) Vas deferens
- (B) Sperm duct
- (C) Prostate gland
- (D) Epididymis
- (E) Urethra

**3:5** Which of the following groups contains the smallest organisms?

- (A) Algae
- (B) Bacteria
- (C) Fungi
- (D) Lichens
- (E) Viruses

**3:6** How many pairs of legs do insects have?

- (A) Two
- (B) Three
- (C) Four
- (D) Five
- (E) Six

**3:7** In the water cycle, which of the following processes **removes** water from the atmosphere?

(A) Evaporation
(B) Photosynthesis
(C) Precipitation
(D) Respiration
(E) Transpiration

**3:8** Which of the following resources could be in limited supply?

I   Food
II   Fossil fuels
III   Mineral resources
IV   Trained human resources

(A) I, II and III only
(B) I, II and IV only
(C) I, III and IV only
(D) II, III and IV only
(E) I, II, III and IV

**3:9** Which of the following characteristics would be shown by all parasites?

(A) Having an insect vector to spread them
(B) Having two hosts
(C) Having hooks and suckers
(D) Producing large numbers of offspring
(E) Killing the host organism

**3:10** Digested food from the small intestine is taken by the blood, first of all, to the:

(A) brain.
(B) liver.
(C) heart.
(D) lungs.
(E) rectum.

**3:11** Which of the following statements are correct?

I   Hormones are transported in the blood plasma.
II   Urea is transported in the blood plasma.
III   Antitoxins are produced by white blood cells.
IV   White blood cells produce toxins against germs.

(A) I, II and III only
(B) II, III and IV only
(C) I, II and IV only
(D) I, III and IV only
(E) I, II, III and IV

The diagram shows the excretory system of a male human. Use the information to answer questions 3:12 and 3:13.

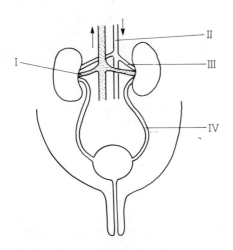

**3:12** The tube labelled IV is called the:

(A) uriniferous tubule.
(B) urethra.
(C) ureter.
(D) glomerulus.
(E) vas deferens.

**3:13** Which structures contain urea, oxygen and food?

(A) I and II
(B) II and III
(C) III and IV
(D) II and IV
(E) I and IV

**3:14** Sweating allows for the removal of which of the following?

I  Heat
II  Water
III  Salt
IV  Carbon dioxide

(A) I, II and III only
(B) II, III and IV only
(C) I and II only
(D) II and III only
(E) III and IV only

**3:15** Which of the following pairs of items is **not** correctly matched?

(A) Ovule – seed
(B) Ovary – fruit
(C) Anther – pollen
(D) Ovule wall – pod
(E) Ovary wall – fruit wall

**3:16** Which of the following is the **most** important feature of prenatal care?

(A) Attending a postnatal clinic
(B) Receiving emotional support from the family
(C) Having a high fluid intake when breast-feeding
(D) Being careful not to trip and fall
(E) Eating a lot of fats and sugars

**3:17** List the characteristics of living organisms, and briefly describe each one. *(9)*

**3:18** The diagram below shows the carbon cycle.

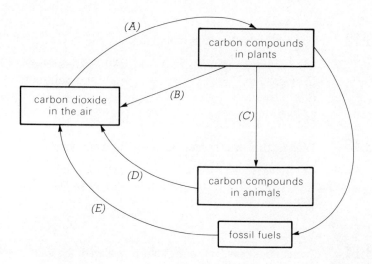

(a) Name each of the processes (A)–(E). *(5)*

(b) Give two ways in which processes (B), (D) and (E) are similar. *(2)*

(c) How does process (A) affect the amount of oxygen in the air? *(1)*

**3:19** (a) Make a table like that below and list two ways in which diagrams I and II differ from each other.

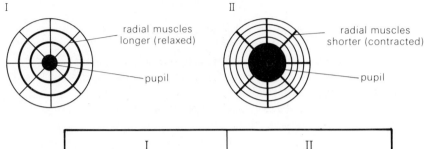

| I | II |
|---|---|
|   |   |
|   |   |

(2)

(b) (i) Under what conditions would the pupil be as in diagram I? (1)
    (ii) What is the advantage of this? (1)

(c) List two ways in which diagrams III and IV differ from each other.

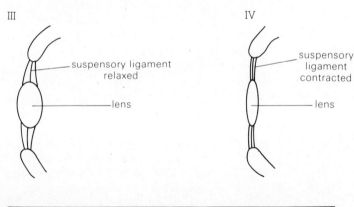

| III | IV |
|---|---|
|   |   |
|   |   |

(2)

(d) (i) Under what conditions would the lens be as in diagram III? (1)
    (ii) Under what conditions would the lens be as in diagram IV? (1)

**3:20** (a) A pure-breeding red-flowered plant was crossed with a pure-breeding white-flowered plant:

$$RR \times rr$$

    (i) What percentage of the gametes made by the red-flowered plant will carry a red allele? *(1)*

    (ii) What percentage of the gametes made by the white-flowered plant will carry a white allele? *(1)*

    (iii) What will be the genotype of the offspring? *(1)*

    (iv) If R is dominant to r, what will be the colour of the offspring? *(1)*

(b) A red-flowered plant (Rr) was crossed with a similar red-flowered plant:

$$Rr \times Rr$$

    (i) What percentage of the gametes produced by these plants will carry:

        (1) a red allele? *(1)*
        (2) a white allele? *(1)*

    (ii) What percentage of the offspring will be:

        (1) red?
        (2) white?

    Show your working. *(3)*

# Test 4

*(Answers on p. 114)*

**4:1** Animals are different from flowering plants because animals:

(A) respond to stimuli.
(B) stay in one place.
(C) move from place to place.
(D) do not need to move.
(E) grow.

**4:2** The group of several species living in a certain place is called the:

(A) habitat.
(B) population.
(C) environment.
(D) community.
(E) ecosystem.

**4:3** Which of the following is **not** a characteristic of leaves which assists photosynthesis? Leaves:

(A) have a large surface area.
(B) are thin and flat.
(C) have waxy cuticle on their epidermis.
(D) have many chloroplasts.
(E) have air spaces between the mesophyll cells.

**4:4** Which of the following is a function of the valves in the veins?

(A) To increase the pressure of the blood
(B) To decrease the pressure of the blood
(C) To stop the backflow of blood
(D) To slow the flow of blood
(E) To produce red blood cells

The diagram on the right shows a section through the elbow joint. Use the information to answer questions 4:5 and 4:6.

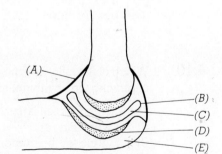

**4:5** On the diagram which label points to the ligament?

**4:6** On the diagram which label points to the space containing synovial fluid?

**4:7** The diagram below shows part of a clinical thermometer. What is the temperature that is shown?

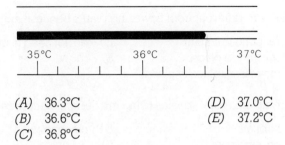

(A) 36.3°C  (D) 37.0°C
(B) 36.6°C  (E) 37.2°C
(C) 36.8°C

**4:8** Seed dispersal is important so that the seedlings:
(A) will all be able to grow on good soil.
(B) will not be overcrowded.
(C) will be pollinated.
(D) will be fertilized.
(E) stay close to the mother plant.

**4:9** Movement in plants could be observed in the:
I   protoplasm.
II  leaves.
III stem.
IV  roots.

(A) I and II only        (D) I, II and III only
(B) II and III only      (E) I, II, III and IV
(C) I and IV only

**4:10** Which of the following things are important in preserving and improving the environment?
I   Setting up fish farms
II  Overfishing and overgrazing
III Caring for the property of others
IV  Using natural resources carefully

(A) I, II and III only    (D) I, III and IV only
(B) II, III and IV only   (E) I, II, III and IV
(C) I, II and IV only

**4:11** Which of the following statements are correct descriptions of a chemical (inorganic) fertilizer?

   I   It releases mineral salts quickly.
   II  It improves the soil texture.
   III It is cheap.
   IV  Its mineral salts are easily leached.

   (A) I and II only
   (B) I and III only
   (C) I and IV only
   (D) II and III only
   (E) III and IV only

**4:12** Which of the following pollutants causes the blackening of buildings and trees?

   (A) Sulphur dioxide
   (B) Soot
   (C) DDT
   (D) Carbon monoxide
   (E) Carbon dioxide

**4:13** Which of the following is **not** a tissue?

   (A) A nerve
   (B) Biceps muscle
   (C) Leaf epidermis
   (D) Leaf mesophyll
   (E) Human egg

**4:14** A de-starched plant is one which:

   (A) is variegated (that is, partly green and partly white).
   (B) cannot make starch.
   (C) has had its starch removed by ethanol.
   (D) has been left in the dark for 24 hours.
   (E) can only make sugars, not starch.

**4:15** Which of the following statements is **not** correct?

   (A) An organism has a constant rate of growth.
   (B) A growing organism needs plenty of protein.
   (C) Growth is a characteristic of all living things.
   (D) Energy is required for growth.
   (E) Growth involves the making of new cells.

**4:16** Which of the following pairs of items, which relate to the menstrual cycle, is **not** correctly matched?

   (A) Days 1–5    — bleeding (menstruation)
   (B) Days 6–12   — repair of the uterus wall
   (C) Days 6–12   — growth of the egg
   (D) Days 13–15  — ovulation
   (E) Days 16–25  — breakdown of the uterus wall

**4:17** The diagram below shows a half-flower of an insect-pollinated flower.

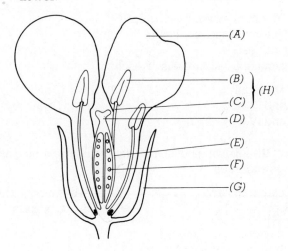

Make a table like that below and write the name of each part and its function.

| Part | Name | Function |
|------|------|----------|
| (A)  |      |          |
| (B)  |      |          |
| (C)  |      |          |
| (D)  |      |          |
| (E)  |      |          |
| (F)  |      |          |
| (G)  |      |          |
| (H)  |      |          |

(8)

**4:18** The table below shows the relative activity of three digestive enzymes at different pH levels.

|         | Relative activity      |                |       |
|---------|------------------------|----------------|-------|
|         | 1                      | 2              | 3     |
| Pepsin  | pH 0.5 and 3.5         | pH 1 and 3     | pH 2  |
| Amylase | pH 5.5 and 8.5         | pH 6 and 8     | pH 7  |
| Lipase  | pH 7.5 and 10.5        | pH 8 and 10    | pH 9  |

Enter the results on the graph paper below, and draw and label three curves to show the readings for the three different enzymes. *(9)*

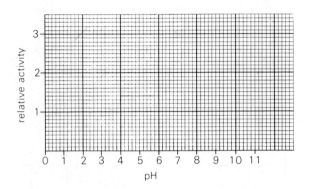

**4:19** The diagram below shows part of a food web in the garden.

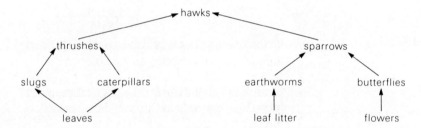

Using the information in the food web:

(a) What is meant by the arrow? *(1)*

(b) What do leaves, leaf litter and flowers have in common? *(1)*

(c) What do slugs, caterpillars, earthworms and butterflies have in common? *(1)*

(d) (i) How do the caterpillars feed? *(1)*
    (ii) How do the butterflies feed? *(1)*

(e) Name a:
    (i) primary consumer. *(1)*
    (ii) secondary consumer. *(1)*
    (iii) tertiary consumer. *(1)*

**4:20** The diagram below is an incomplete diagram of the apparatus used to compare the carbon dioxide concentration in inhaled and exhaled air.

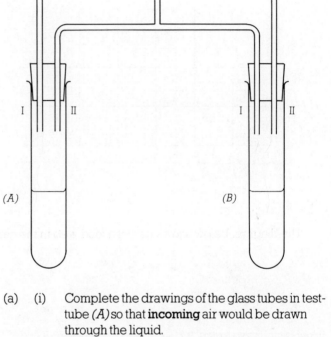

(a) (i) Complete the drawings of the glass tubes in test-tube *(A)* so that **incoming** air would be drawn through the liquid. *(1)*
    (ii) Complete the drawings of the glass tubes in test-tube *(B)* so that **exhaled** air would be passed through the liquid. *(1)*

(b) (i) What should be the liquid in test-tube *(A)*? *(1)*
    (ii) What will happen to it, and why? *(1)*
    (iii) What is the reason for having test-tube *(A)*? *(1)*

(c) (i) What should be the liquid in test-tube *(B)*? *(1)*
    (ii) What will happen to it, and why? *(1)*

(d) (i) What is the experiment set up to show? *(1)*
    (ii) Do you think it is a fair test? Explain your answer. *(1)*

# Test 5

*(Answers on p. 116)*

**5:1** Which of the following statements concerned with the transfer of energy is **correct**? Energy from the sun:

(A) can be recycled in nature.
(B) is used by plants to make food which is eaten by animals to release energy for life processes.
(C) is used by animals to make food which is used by other organisms.
(D) is used by decomposers to produce organic compounds used by plants and animals to release energy.
(E) is used directly by heterotrophs.

**5:2** Disease can cause:

I ill-health and loss of efficiency in humans.
II death of humans.
III loss of farm animals.
IV reduction in agricultural production.

(A) I, II and III only
(B) I, II and IV only
(C) I, III and IV only
(D) II, III and IV only
(E) I, II, III and IV

**5:3** Which of the following substances releases the most energy when combusted with oxygen?

(A) 1 g of amino acid
(B) 1 g of starch
(C) 1 g of glucose
(D) 1 g of sucrose
(E) 1 g of fat

**5:4** Woodlice:

(A) move towards damp, bright places.
(B) move towards damp, dark places.
(C) move towards dry, bright places.
(D) are unaffected by light intensity.
(E) are unaffected by variations in moisture.

**5:5** How do human males and females differ in their sex chromosomes?

(A) Males have an X chromosome and females a Y chromosome.
(B) Males have a Y chromosome and females an X chromosome.
(C) Males have XX chromosomes and females XY chromosomes.
(D) Males have XY chromosomes and females XX chromosomes.
(E) Males have XY chromosomes and females XXX chromosomes.

**5:6** Organisms can be classified using:

(A) keys.
(B) quadrats.
(C) sieves.
(D) traps.
(E) funnels.

**5:7** A community is:

(A) a place where organisms live.
(B) a group of organisms of the same species.
(C) a group of several species living together.
(D) the conditions in a particular habitat.
(E) the interactions in a particular habitat.

**5:8** Part of a food web is shown below.

```
              Aphids → Ladybirds → Sparrows
Leaves ↗                              ↓
       ↘     Slugs →  Thrushes →    Hawks
```

Which of the organisms are secondary consumers?

(A) Aphids and slugs
(B) Thrushes and sparrows
(C) Thrushes and ladybirds
(D) Sparrows and hawks
(E) Sparrows and ladybirds

**5:9** 50 g of soil was heated to constant mass to drive off the water which was in it. When the soil was cooled its mass was 40 g. The percentage of water originally in the soil was:

(A) 10 per cent.
(B) 20 per cent.
(C) 30 per cent.
(D) 40 per cent.
(E) 50 per cent.

**5:10** The vacuoles of plant cells contain:

I   chloroplasts.
II  water.
III salts.
IV  starch grains.

(A) I, II and III only
(B) II, III and IV only
(C) II and III only.
(D) II and IV only
(E) I, II, III and IV

**5:11** Which of the following are characteristics of red blood cells in mammals?

I   Contain haemoglobin
II  Have a lobed nucleus
III Made in the red bone marrow
IV  Destroy invading micro-organisms

(A) I and II only
(B) II and III only
(C) II and IV only
(D) I and III only
(E) I, II and III only

Use the information in the table below to answer questions 5:12 and 5:13.

| Organ | Blood flow (cm³/100 g of organ) |
|---|---|
| (A) | 150 |
| (B) | 70 |
| (C) | 25 |
| (D) | 20 |
| (E) | 1 |

(A), (B), (C), (D) and (E) are five organs in the human body with their blood flow shown when at rest. They are the small intestine, large intestine, liver, leg, and stomach (but **not** in that order).

**5:12** Which is the **most** metabolically active organ?

**5:13** Which organ is **most** likely to be the leg?

**5:14** Urine is made mainly of:

(A) water.
(B) urea.
(C) sugar.
(D) dissolved salts.
(E) uric acid.

**5:15** Which of the following statements about the amniotic fluid are correct?

I   It provides the foetus with oxygen, food and antibodies.
II  It keeps the foetus from drying out.
III It keeps an equal pressure on the foetus.
IV  It cushions the foetus from mechanical shock.

(A) I, II and III only
(B) I, II and IV only
(C) I, III and IV only
(D) II, III and IV only
(E) I, II, III and IV

**5:16** Mitosis is important because:

(A) it halves the chromosome number.
(B) it maintains the chromosome number.
(C) it is the only way that plant cells divide.
(D) it is the only way that animal cells divide.
(E) it occurs in the production of gametes.

**5:17** (a) Describe three differences between living and non-living things. *(3)*

(b) Describe three differences between plants and animals. *(6)*

**5:18** The table below shows some of the organisms in a certain habitat, and the organisms which feed on them.

| Organisms which are eaten | Organisms that eat them |
|---|---|
| Grass | Cows and grasshoppers |
| Cabbages | Slugs |
| Decaying plants | Earthworms |
| Flowers (nectar) | Butterflies |
| Cows | Humans |
| Grasshoppers | Frogs |
| Slugs, earthworms and butterflies | Small birds |
| Frogs and small birds | Large birds |

(a) Use the information from the table to identify four food chains each containing at least three organisms. Write out the four food chains you choose. *(4)*

(b) Design and draw a food web to represent the feeding relationships of **all** the organisms in the table. *(5)*

**5:19** (a) The following statements describe steps in the process of pollination and fertilization. Rewrite the numbers of the statements in their correct order.

    (i)   Pollen tubes are formed.
    (ii)  Pollen grains settle on the stigma.
    (iii) The nucleus from the pollen tube fuses with the egg cell nucleus.
    (iv) Pollen tubes grow down the style.
    (v)  Ripe pollen is shed from an anther. *(4)*

(b) Look at the tables below. Match the words on the left with the correct statements on the right to describe what happens to them after fertilization. Draw lines to show which ones go together. *(4)*

| Structure | | After fertilization: | |
|---|---|---|---|
| (A) | Stamens | (i) | becomes a seed. |
| (B) | Ovule | (ii) | wither and die. |
| (C) | Petals | (iii) | becomes the fruit wall. |
| (D) | Ovary | (iv) | becomes a fruit |
| (E) | Ovary wall | (v) | wither and die. |

**5:20** The diagram below shows the reflex arc involved when a person touches a hot object and quickly lifts his or her hand away.

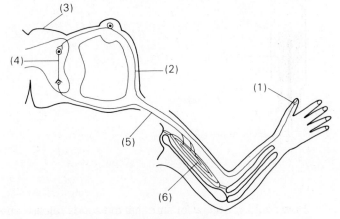

(a) Name the parts labelled (1)–(6) *(3)*

(b) Describe the parts played by (2), (4), (5) in a reflex arc. *(3)*

(c) What happens to structure (6) when it receives impulses from the part labelled (5)? *(1)*

(d) Of what value is this kind of reaction to the organism? *(1)*

# Test 6

*(Answers on p. 117)*

**6:1** Which group of vertebrates is cold blooded and has scales and fins?

(A) Fish
(B) Amphibia
(C) Reptiles
(D) Birds
(E) Mammals

**6:2** Decomposers:

(A) eat living plants.
(B) feed on organic materials.
(C) live inside animals and parasitize them.
(D) are autotrophic.
(E) do not need energy to live.

**6:3** The apparatus shown below is used to demonstrate the release of energy by germinating seeds.

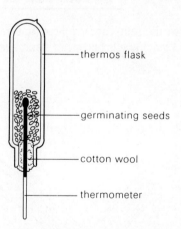

How could you show for sure that the seeds release energy during germination? By repeating the experiment, and putting in the thermos flask:

(A) another set of germinating seeds to check your results.
(B) some dead seeds.
(C) twice as many germinating seeds as before.
(D) some dead things such as stones.
(E) some small live animals.

**6:4** Which of the following is an adaptation favouring wind pollination?

(A) Long feathery stigmas
(B) Scent
(C) Nectaries
(D) Large coloured petals
(E) Short filaments

**6:5** A unicellular organism, e.g. *Amoeba*, is different from more complex organisms because it does **not**:

(A) respond to stimuli.
(B) reproduce.
(C) contain tissues.
(D) excrete.
(E) contain chromosomes.

**6:6** The place where a particular organism is found is called its:

(A) physical environment.
(B) biological environment.
(C) niche.
(D) habitat.
(E) community.

**6:7** A quantity of pesticide was accidentally spilt into a river. After a period of time which of the aquatic organisms would contain the highest concentration of the pesticide?

(A) Algae
(B) Pondweed
(C) Tadpoles
(D) Small fish
(E) Large fish

**6:8** In a certain habitat hawks fed on thrushes, which fed on grasshoppers, which fed on grass. The thrushes could be described as:

I   predators.
II  prey.
III primary consumers.
IV  secondary consumers.

(A) II and IV only
(B) I, II and III only
(C) I, II and IV only
(D) I, III and IV only
(E) I, II, III and IV

**6:9** Which of the following comparisons of plant and animal cells is **not** correct?

| | *Typical plant cells* | *Typical animal cells* |
|---|---|---|
| (A) | Cellulose cell wall | No cell wall |
| (B) | Cell membrane | No cell membrane |
| (C) | Chloroplasts | No chloroplasts |
| (D) | Starch grains | No starch grains |
| (E) | Large vacuoles | Small vacuoles |

**6:10** Which of the following is **not** needed for photosynthesis to occur?

(A) Light
(B) Water
(C) Oxygen
(D) Carbon dioxide
(E) Chlorophyll

**6:11** Under which of the following weather conditions would transpiration be **most** rapid?

(A) Hot, dry and windy
(B) Hot, humid and still
(C) Hot, dry and still
(D) Cool, dry and windy
(E) Cool, humid and still

**6:12** It has been suggested that our diet should contain:

I   less saturated fat
II  less sugar.
III more fruit and vegetables.
IV  more grains.

This is because these foods have an effect on our:

(A) circulatory system only.
(B) digestive system only.
(C) respiratory system only.
(D) circulatory and digestive systems only.
(E) circulatory, digestive and respiratory systems.

Use the diagram below, which shows a simple reflex arc, to answer questions 6:13 and 6:14.

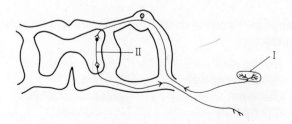

**6:13** Label I is the:

(A) effector.
(B) motor nerve fibre.
(C) spinal nerve
(D) white matter.
(E) receptor.

**6:14** Label II is the:

(A) motor neurone.
(B) cell body of the sensory neurone.
(C) connecting neurone in the spinal cord.
(D) primary receptor in the spinal nerve.
(E) effector in the spinal nerve.

Use the diagram below of the human life cycle to answer questions 6:15 and 6:16.

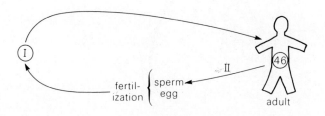

**6:15** What would be the chromosome number in cell I?

(A) 22
(B) 23
(C) 23–46
(D) 46
(E) 92

**6:16** What kind of cell division occurs at position II?

(A) Mitosis
(B) Meiosis
(C) Mitosis to produce eggs, meiosis to produce sperm
(D) Meiosis to produce eggs, mitosis to produce sperm
(E) Ordinary cell division

**6:17** The diagram below shows the main parts of the human respiratory system.

(a) Name the parts labelled *(A)* to *(J)*. *(5)*
(b) Which part closes during swallowing? *(1)*
(c) In which part does gaseous exchange occur? *(1)*
(d) Which part is raised during exhalation (breathing out)? *(1)*

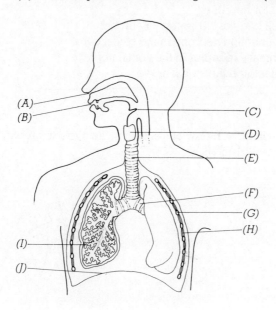

**6:18** The following apparatus was used to find out if carbon dioxide was necessary for photosynthesis.

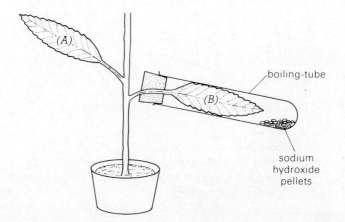

(a) (i) Before the experiment begins, what should be done to the whole plant? *(1)*
(ii) What is the reason for doing this? *(1)*

(b) (i) When the experiment begins, where should the plant be left? Why? *(1)*
(ii) What is the purpose of the sodium hydroxide pellets in the boiling tube near to leaf B? *(1)*

(c) When removed from the plant, leaves A and B were tested for starch. Give the correct sequence for doing this from the steps listed below. *(2)*

    (i) Boil in alcohol (ethanol).
    (ii) Dip in boiling water.
    (iii) Test with iodine solution.
    (iv) Wash in water.

(d) (i) Look again at the diagram of the apparatus. How could the experiment be improved? *(1)*
(ii) Why is the improvement necessary? *(1)*

**6:19** The diagram below shows the nitrogen cycle.

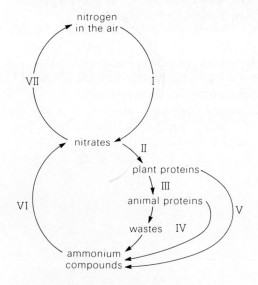

Use the diagram to explain how:

(a) nitrates are used up, and *(4)*
(b) nitrates are produced. *(5)*

(In your answer refer to the numbered stages I to VII.)

**6:20** Examine the graph below, which shows the growth of boys and girls from 10 to 18 years.

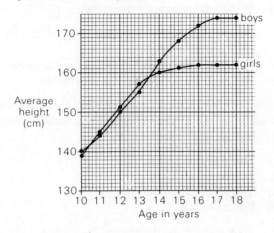

By reference to the graph, write down which of the following statements are true and which are false.

- (A) Boys are always taller than girls.
- (B) Girls have their spurt of fast growth before boys.
- (C) Boys stop growing before girls.
- (D) There is a constant rate of growth between the 13th and 15th years.
- (E) Boys have their greatest rate of growth in their 14th year.
- (F) Boys grow at a steady rate in their 11th, 12th and 13th years.
- (G) Girls stop growing at about 16 years.
- (H) Girls grow at 6 cm/year in their 11th, 12th and 13th years.
- (I) Girls of 11 and 12 years are on average taller than boys of the same age. (9)

# Test 7

*(Answers on p. 119)*

**7:1** Which group of vertebrates is warm blooded, has four limbs and hair?

(A) Fish
(B) Amphibia
(C) Reptiles
(D) Birds
(E) Mammals

**7:2** The conditions which exist in a particular habitat are called the:

(A) physical environment.
(B) biotic environment.
(C) physical and biotic environment.
(D) community.
(E) population.

**7:3** Which of the following are characteristics of enzymes?

I   They can work only at body temperature.
II  They are made only in the gut.
III They act only on particular substrates.
IV  They work only at particular ranges of pH.

(A) I and II only
(B) III and IV only
(C) I and III only
(D) II and IV only
(E) I, II and IV only

**7:4** What do the following structures have in common?

I   Cell membrane of unicellular organism
II  Alveoli
III Intercellular leaf spaces

(A) Sensitivity
(B) Absorbing salts
(C) Transpiration
(D) Osmoregulation
(E) Gaseous exchange

**7:5** Which of the following is the **most** important feature of postnatal care?

(A) Attending a prenatal clinic
(B) Breast-feeding a baby
(C) Visiting the dentist
(D) Not smoking, or drinking alcohol
(E) Being careful not to trip or fall

**7:6** Fungi, such as mould, feed:

- (A) like plants.
- (B) like *Amoeba*.
- (C) on inorganic food.
- (D) by chemosynthesis.
- (E) by using digestive enzymes.

**7:7** A quadrat would be useful for:

- (A) measuring angles.
- (B) sieving soil.
- (C) sampling plants or animals.
- (D) investigating particle size.
- (E) collecting small animals.

**7:8** Which of the following **could** be a secondary consumer?

- I   Carnivore
- II  Herbivore
- III Omnivore
- IV  Parasite

- (A) I and II only
- (B) I and III only
- (C) I, II and III only
- (D) I, III and IV only
- (E) II, III and IV only

**7:9** Which of the following is biodegradable?

- (A) Iron
- (B) Aluminium
- (C) Glass
- (D) Paper
- (E) Plastic

Use the diagram below to answer questions 7:10 and 7:11.

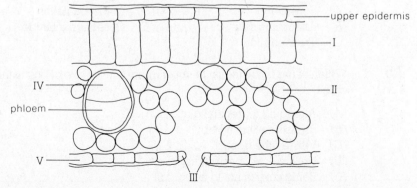

**7:10** Which tissue contains the most chloroplasts?

(A) I   (B) II   (C) III   (D) IV   (E) V

**7:11** The function of tissue IV is to:

(A) take food away from the leaf.
(B) store food.
(C) bring water to the leaf.
(D) allow water vapour to escape.
(E) allow for gaseous exchange.

**7:12** Which of the following enzymes is concerned with the digestion of starch?

(A) Lipase          (D) Protease
(B) Amylase         (E) Trypsin
(C) Pepsin

Use the diagram of a section of the mammalian eye below to answer questions 7:13 and 7:14.

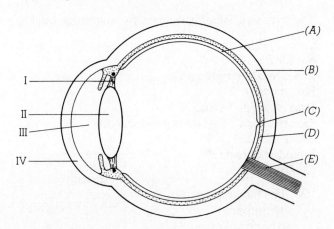

**7:13** In the diagram which of the structures (A)–(E) is the choroid?

**7:14** Which of the structures labelled refract light rays?

(A) I and II only       (D) II only
(B) II and III only     (E) III only
(C) II and IV only

Use the information below to answer questions 7:15 and 7:16.

Ten laboratory animals were kept in a cage and were measured from the time that they hatched until day 14. The average length of the animals on certain days is given in the table below.

| Age in days | 1 | 2 | 3 | 5 | 6 | 8 | 9 | 10 | 12 | 13 | 14 |
|---|---|---|---|---|---|---|---|---|---|---|---|
| Average length (mm) | 3 | 3 | 5 | 5 | 8 | 8 | 8 | 12 | 12 | 12 | 17 |

**7:15** What is happening on days 6 to 9 and on days 10 to 13? The animals are:

(A) growing rapidly.
(B) growing slowly.
(C) at constant length.
(D) hatching.
(E) metamorphosing.

**7:16** The laboratory animals could have been:

(A) cockroaches.
(B) mice.
(C) tadpoles.
(D) lizards.
(E) fish.

**7:17** The sources of sugar consumed by an average adult are given below:

Sweets 2%
Bread and other grains 6%
Baked goods 13%
Jams, sugar and syrup 18%
Ice cream and other dairy products 10%
Soft drinks 25%
Breakfast cereals 5%
Other foods 21%

(a) Use the graph paper to prepare a bar chart of this information. *(8)*

(b) Which group accounts for a quarter of the sugar intake? *(1)*

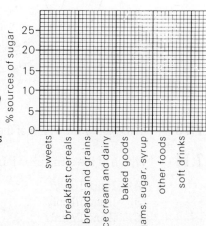

**7:18** (a) Give five ways in which humans have polluted the environment, and in each case say how animals or plants have been affected. *(5)*

(b) Give four ways in which humans can conserve the environment. *(4)*

**7:19** The diagram below shows the markings on the hard fore-wings of five different ladybirds.

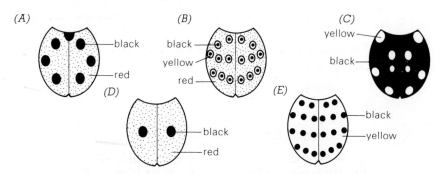

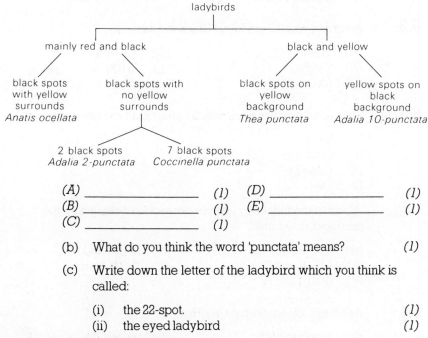

(a) Use the key below to identify the ladybirds.

(A) _____ *(1)*  (D) _____ *(1)*
(B) _____ *(1)*  (E) _____ *(1)*
(C) _____ *(1)*

(b) What do you think the word 'punctata' means? *(1)*

(c) Write down the letter of the ladybird which you think is called:

   (i) the 22-spot. *(1)*
   (ii) the eyed ladybird *(1)*

**7:20** Design a series of experiments to test the hypothesis: 'Large seeds **germinate** faster than small seeds.' *(8)*

# Test 8

*(Answers on p. 120)*

**8:1** When humans go up in a spacecraft they have to take with them a supply of:

    I    oxygen.
    II   food.
    III  water.
    IV  carbon dioxide.

(A) I, II and III only
(B) II, III and IV only
(C) I, II and IV only
(D) I, III and IV only
(E) I, II, III and IV

**8:2** An insect was 0.5 cm long. This is the same as:

(A) 5 mm.
(B) 50 mm.
(C) 500 mm.
(D) 5000 nm.
(E) 50 000 nm.

**8:3** Here are four organisms which make up a food chain:

    I    Snail
    II   Cabbage
    III  Small bird
    IV  Owl

Which of the following lists is a correct food chain with the producer listed first?

(A) I, II, III and IV
(B) I, III, II and IV
(C) II, III, I and IV
(D) II, I, III and IV
(E) IV, III, II and I

**8:4** An acidic soil would have a pH of:

(A) less than pH 7.
(B) pH 7.
(C) more than pH 7.
(D) anything from pH 5–8.
(E) anything from pH 8–14.

**8:5** Photosynthesis occurs in the:

- (A) starch grains.
- (B) mitochondria.
- (C) phloem.
- (D) chloroplasts.
- (E) chromosomes.

**8:6** Which of the following substances would give a positive test with Benedict's solution?

- (A) Cellulose
- (B) Starch
- (C) Fat
- (D) Protein
- (E) Glucose

**8:7** Blood travels away from the heart in:

- (A) capillaries.
- (B) veins.
- (C) arteries.
- (D) valves.
- (E) lymph vessels.

**8:8** Which of the following pairs of items is **not** correct?

|     | *Aerobic respiration* | *Photosynthesis* |
| --- | --- | --- |
| (A) | Energy is released. | Energy is needed. |
| (B) | Oxygen is used up. | Oxygen is produced. |
| (C) | Water is produced. | Water is needed. |
| (D) | Carbon dioxide is produced. | Carbon dioxide is needed. |
| (E) | Occurs during the night only. | Occurs during the day only. |

**8:9** Which of the following statements is **correct**?

- (A) Responses are the nervous impulses from sense organs.
- (B) Stimuli are changes in the internal or external environments.
- (C) Receptors are the muscles and glands which respond to stimuli.
- (D) Effectors are the sensory cells and organs receiving stimuli.
- (E) Effectors have to be stimulated before nervous impulses are set up.

**8:10** Which of the following are functions of the human skeleton?

- I  Locomotion
- II  Support
- III  Protection
- IV  Blood cell production

- (A) I and II only
- (B) II and III only
- (C) I, II and III only
- (D) I, III and IV only
- (E) I, II, III and IV

**8:11** Arthropods can be distinguished from all other invertebrates because they have:

(A) antennae.
(B) segmented bodies.
(C) no backbone.
(D) a hard covering.
(E) jointed legs.

**8:12** Which of the following correctly lists the processes, in the right order, for the transfer of energy from the sun to that used by animals for movement?

(A) Photosynthesis→food storage→absorption→respiration
(B) Photosynthesis→food storage→feeding→digestion →absorption→respiration
(C) Photosynthesis→digestion→food storage→absorption →respiration
(D) Feeding→food storage→absorption→digestion →respiration
(E) Photosynthesis→digestion→feeding→respiration

**8:13** The hormone adrenaline has which of the following effects on the body?

I   Breathing rate increases.
II  More blood goes to the skin.
III More blood goes to the muscles.
IV  Heart beats faster.

(A) I and IV only
(B) I, II and III only
(C) I, III and IV only
(D) I, II and IV only
(E) I, II, III and IV

**8:14** Which of the following diagrams is the **best** representation of a flowering plant life cycle?

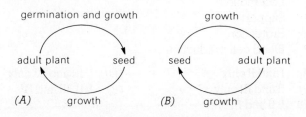

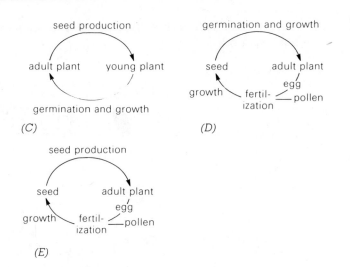

**8:15** What is the **best** estimate of the surface area of the leaf shown below?

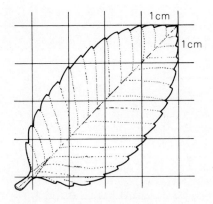

(A) 1 cm²   (B) 2 cm²   (C) 10 cm²   (D) 12 cm²   (E) 14 cm²

**8:16** A red-flowered plant was self-fertilized. When its seeds were grown the plants produced flowers, some of which were white. The **most** likely explanation was that the original red-flowered plant:

(A) was homozygous.
(B) was heterozygous.
(C) was variegated.
(D) had a deficiency disease.
(E) had a gene mutation.

**8:17** The diagram below shows five insect larvae. The head end is on the right-hand side in each case.

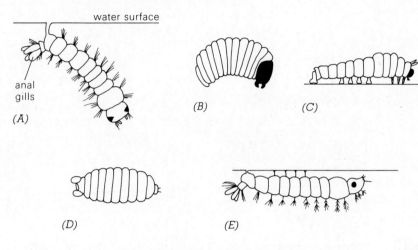

(a) Use the key below to identify the insect larvae.

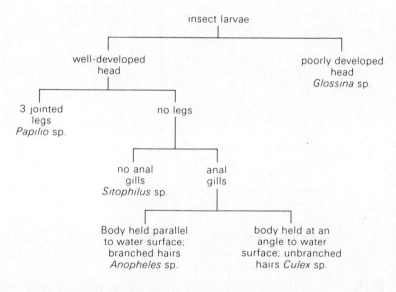

What are the names of each of the larvae?

(A) _____ (1)
(B) _____ (1)
(C) _____ (1)
(D) _____ (1)
(E) _____ (1)

(b) (i) Which of the larvae shown are aquatic? *(1)*
(ii) What structures would assist them in gaseous exchange? *(1)*

(c) (i) Which of the larvae shown would you expect to find feeding on leaves? *(1)*
(ii) What is your reason for your choice? *(1)*

**8:18** The constituents of a slimmer's dinner are given below.

|  | Per 100 g | Per 300 g serving |
|---|---|---|
| kilojoules | 341 | |
| protein | 6.8 g | |
| carbohydrate | 9.0 g | |
| fat | 2.2 g | |

(a) Complete the figures on the right-hand side of the table to show the amounts of each of the constituents in the whole serving. *(2)*

(b) (i) What is the percentage of protein in 100 g? *(1)*
(ii) What is the percentage of protein in 300 g? *(1)*

(c) The ingredients of the meal are given below.

---
INGREDIENTS: ZUCCHINI, TOMATO PUREE, TOMATOES, COTTAGE CHEESE, MOZZARELLA CHEESE, PASTA, ONIONS, CARROTS, MODIFIED STARCH, SALT, WHEAT FLOUR, BRAN, SUGAR, HERBS, HYDROLYSED VEGETABLE PROTEIN, PEPPER.
THIS PRODUCT IS MEAT FREE.

This dish provides approx. 30% of the protein recommended daily for most people.

---

(i) The dish is meat-free, but provides approximately 30 per cent of the daily protein needs. How is this possible? *(2)*
(ii) Apart from its low energy value give two ways in which this dish could form part of a healthy diet. *(2)*

**8:19** The percentage of the total population of the UK in different age groups is given below for the years 1850, 1900 and 1950, and an estimate for the year 2000.

|  | % under 20 | % between 20 and 65 | % over 65 |
|---|---|---|---|
| 1850 | 45 | 50 | 5 |
| 1900 | 42 | 48 | 10 |
| 1950 | 33 | 55 | 12 |
| 2000 | 25 | 50 | 25 |

(a) Show this information on the bar chart below.
The information for 1850 has been done.
Use the same shading for the other columns *(6)*

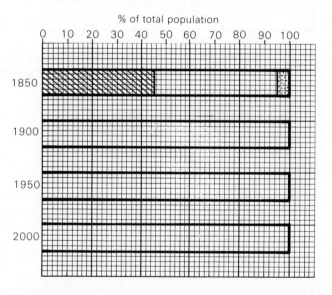

(b) Complete the key given below.

*(1)*

(c) Give two reasons why the percentage of people over 65 years more than doubled from 1850 to 1950. *(2)*

**8:20** The diagram below shows the human foetus in the uterus.

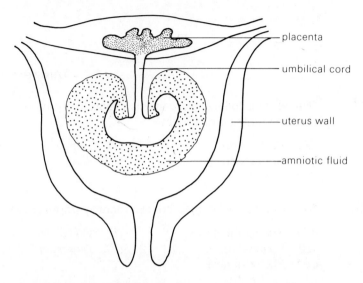

Write an account of how the foetus is cared for in the uterus. Include the functions of the parts labelled in the diagram. *(9)*

# Test 9

*(Answers on p. 122)*

**9:1** Which of the following organisms is **most** likely to be harmed by the drying effects of the sun?

(A) Earthworm
(B) Cockroach
(C) Lizard
(D) Human
(E) Snail

**9:2** Which of the following is the basic energy-trapping life process?

(A) Transpiration
(B) Respiration
(C) Photosynthesis
(D) Cell division
(E) Digestion

**9:3** In the nitrogen cycle which of the following terms best describes the change of dead organisms and faeces into ammonium compounds?

(A) Fermentation
(B) Decay
(C) Nitrification
(D) Nitrogen fixing
(E) Nutrition

**9:4** Which of the following can be caused by pollution of the environment?

I   Death of aquatic organisms
II  Formation of cancers
III Respiratory diseases
IV  Deterioration of buildings

(A) I, II and III only
(B) II, III and IV only
(C) I, II and IV only
(D) I, III and IV only
(E) I, II, III and IV

**9:5** Which of the following is **not** caused by burning fossil fuels?

(A) Destruction of forests
(B) Respiratory diseases
(C) Blackening of buildings
(D) Formation of smog
(E) Using up natural resources

**9:6** What happens when plant cells are left in distilled water? The cells:

(A) swell up and burst.
(B) excrete the water through their vacuoles.
(C) become turgid.
(D) become plasmolysed.
(E) die.

**9:7** Which class of foods contains carbon, nitrogen, oxygen and hydrogen?

(A) Starches
(B) Sugars
(C) Saturated fats
(D) Unsaturated fats
(E) Proteins

**9:8** Which of the following statements is **correct**?

(A) The living xylem vessels transport water and salts.
(B) The dead xylem vessels transport food substances.
(C) The dead phloem sieve tubes transport food substances.
(D) The living phloem sieve tubes transport food substances.
(E) The living phloem sieve tubes transport water and food.

**9:9** The alveolus of a mammal is shown below, in contact with a blood capillary.

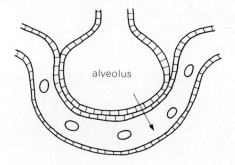

What is the **main** substance which travels along the arrow marked on the diagram?

(A) Exhaled air
(B) Oxygen
(C) Carbon dioxide
(D) Oxygenated blood
(E) Deoxygenated blood

**9:10** In which part of the flower is the female gamete made?

(A) Stigma  (D) Ovary
(B) Style   (E) Filament
(C) Ovule

**9:11** The brown eye allele (B) is dominant to the blue eye allele (b). What genotypes correspond to a phenotype of brown eyes?

(A) BB only            (D) BB and Bb only
(B) Bb only            (E) BB, Bb and bb only
(C) Bb and bb only

**9:12** A person who is working underwater for long periods of time would need which of the following items?

I   Food
II  Water
III Air
IV  Light

(A) I and II only       (D) II, III and IV only
(B) I and III only      (E) I, II, III and IV
(C) I, II and III only

**9:13** Which of the following statements is **correct**?

(A) Antibiotics and antitoxins are drugs used against germs.
(B) Antitoxins are produced by white cells.
(C) Antibodies are produced by red cells.
(D) Antibiotics kill all pathogenic organisms.
(E) Toxins kill all pathogenic organisms.

**9:14** Which of the following substances has the highest energy content per gram (kJ/g)?

(A) Starch     (D) Fish
(B) Sugar      (E) Margarine
(C) Cheese

**9:15** From the liver the blood goes along which of these pathways?

(A) Small intestine→heart→lungs→body cells
(B) Heart→lungs→heart→body cells
(C) Lungs→heart→body cells→heart
(D) Heart→lungs→body cells→heart
(E) Lungs→heart→kidneys→body cells

**9:16** Which of the following pairs of items is **not** correctly matched?

(A) Scrotum — sac of skin surrounding the testes
(B) Sperm ducts — coiled tubes inside the testes
(C) Prostate gland — produces seminal fluid
(D) Penis — surrounds the urethra
(E) Testes — produces sperm

**9:17** (a) The figure below shows the markings on five different kinds of young salmon. Use the key below to identify each of the fish (A) to (E).

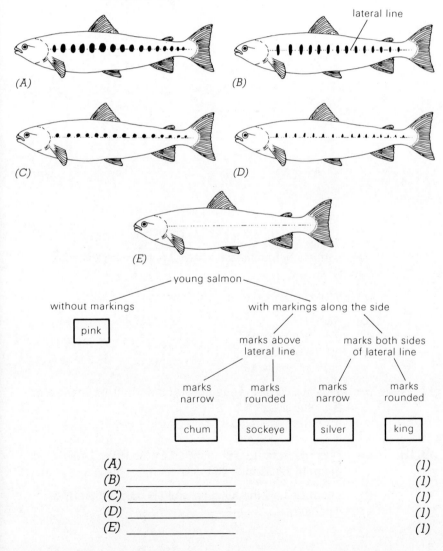

(A) _____ (1)
(B) _____ (1)
(C) _____ (1)
(D) _____ (1)
(E) _____ (1)

(b) Which of the following statements about fish are true, and which are false?

(i) Fish are invertebrates.
(ii) Fish have an internal skeleton.
(iii) Among living things only fish have fins.
(iv) Fish are the only animals that can live underwater. *(4)*

**9:18** The figure below shows part of a food web on waste ground.

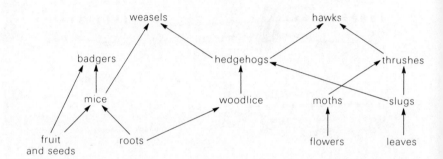

Use information from the food web to answer the following questions.

(a) Write out a food chain which has four organisms in it. *(1)*

(b) Name:
(i) a herbivore. *(1)*
(ii) a predator. *(1)*
(iii) an omnivore. *(1)*

(c) What would be the effect of the death of all the thrushes on:
(i) the slugs? *(2)*
(ii) the hedgehogs? *(2)*

**9:19** (a) Describe how urea is excreted by the body. Draw a diagram to illustrate your answer. *(5)*

(b) Describe how this function could be taken over by a machine. *(3)*

**9:20** The number of chromosomes in a particular mammal and flowering plant are shown below, before and after certain processes have occurred.

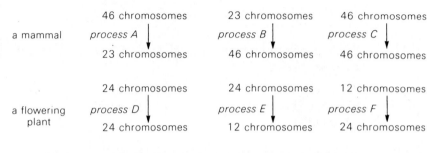

(a) Name the processes B, C, E and F, and for each one name a place in the organism where it may occur. *(4)*

(b) For the particular mammal described could there be a process for 23 chromosomes→23 chromosomes? Explain your answer. *(1)*

(c) Explain why each of the following generalizations is or is not correct.

(i) If an organism reproduces by sexual reproduction it is necessary to have meiosis in the life cycle. *(1)*

(ii) If an organism reproduces only by asexual reproduction it is not necessary to have meiosis in the life cycle. *(1)*

(iii) Animals always have more chromosomes than plants. *(1)*

(iv) Similar processes occur in the life cycle of a mammal and flowering plant. *(1)*

# Test 10

*(Answers on p. 124)*

**10:1** Which group of vertebrates has four limbs, is cold blooded and has a smooth skin?

(A) Fish
(B) Amphibia
(C) Reptiles
(D) Birds
(E) Mammals

**10:2** To which of the following groups does a horsechestnut tree belong?

(A) Algae
(B) Ferns
(C) Conifers
(D) Flowering plants
(E) Monocotyledons

**10:3** Spraying with pesticides may lead to the presence of harmful chemicals in:

I  river water.
II  the atmosphere.
III  animal flesh.
IV  plant storage organs.

(A) I, II and III only
(B) II, III and IV only
(C) I, II and IV only
(D) I, III and IV only
(E) I, II, III and IV

**10:4** A person can become immune to a certain disease by producing:

(A) antibodies.
(B) antibiotics.
(C) antigens.
(D) toxins.
(E) pathogens.

**10:5** Which of the following diagrams shows a longitudinal section of a dicotyledonous stem?

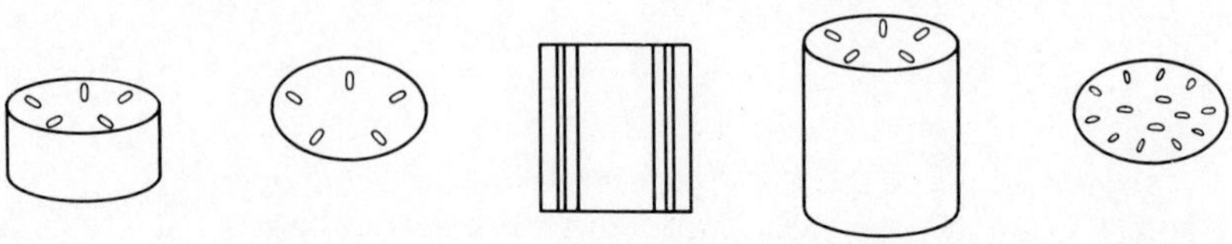

(A)   (B)   (C)   (D)   (E)

**10:6** Deficiency diseases occur:

(A) only in plants growing in the soil.
(B) only in plants growing in water cultures.
(C) only in plants growing in sand cultures.
(D) only in animals.
(E) in plants and animals.

**10:7** In which of the following places would you expect the nerve endings sensitive to touch to be the farthest apart?

(A) Finger tips
(B) Palm of the hand
(C) Back of the hand
(D) Upper arm
(E) Soles of the feet

The diagram on the right shows a section through the mammalian skin. Use the information to answer questions 10:8 and 10:9.

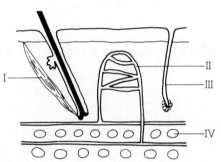

**10:8** Which structures are concerned with reducing the temperature of the body?

(A) I and II only
(B) II and III only
(C) III and IV only
(D) I, II and IV only
(E) I, II, III and IV

**10:9** Which of the following processes is involved in reducing body temperature?

(A) Contraction
(B) Translocation
(C) Evaporation
(D) Respiration
(E) Condensation

**10:10** In order to grow a seedling needs:

(A) oxygen and light only.
(B) oxygen, light and water only.
(C) carbon dioxide and water only.
(D) oxygen, light, water and mineral salts only.
(E) carbon dioxide, oxygen, light, water and mineral salts only.

**10:11** Why are offspring of the same parents different from each other? Because:

(A) meiosis occurs during the formation of gametes.
(B) mitosis occurs during the formation of gametes.
(C) the offspring are produced at different times.
(D) the offspring are of different ages.
(E) the offspring develop differently in the uterus.

**10:12** Which of the following statements is **not** a correct description of a consumer? Consumers are:

(A) animals.
(B) heterotrophic.
(C) autotrophic.
(D) mobile.
(E) holozoic.

**10:13** Denitrifying bacteria change:

(A) nitrates to nitrogen.
(B) nitrogen to nitrates.
(C) nitrites to nitrates.
(D) ammonia to nitrites.
(E) body wastes to ammonia.

**10:14** Which of the following plant structures could provide food for animals?

I   Roots
II  Leaves
III Flowers
IV  Seeds

(A) I, II and III only
(B) II, III and IV only
(C) I, II and IV only
(D) I, III and IV only
(E) I, II, III and IV

**10:15** Which of the following statements about digestion is **not** correct? Digestion:

(A) changes starch to glucose.
(B) changes proteins to amino acids.
(C) releases energy for the body.
(D) makes food particles smaller.
(E) makes food particles soluble.

**10:16** Which of the following processes does **not** depend upon the principles of cloning (asexual reproduction)?

  (A) Budding of yeast
  (B) Binary fission
  (C) Formation of gametes
  (D) Multiple fission
  (E) Vegetative reproduction

**10:17** (a) Use the key provided to identify the fish *(A), (B), (C), (D)* and *(E)* shown below. The key can be used for six fish, but you are asked to identify only the five that are shown. Write the letter of each animal in the box close to its name. *(5)*

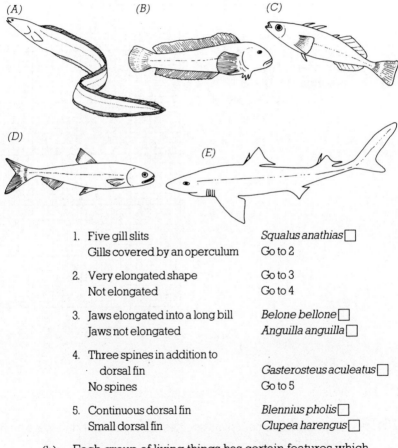

| | | |
|---|---|---|
| 1. Five gill slits | *Squalus anathias* ☐ | |
| Gills covered by an operculum | Go to 2 | |
| 2. Very elongated shape | Go to 3 | |
| Not elongated | Go to 4 | |
| 3. Jaws elongated into a long bill | *Belone bellone* ☐ | |
| Jaws not elongated | *Anguilla anguilla* ☐ | |
| 4. Three spines in addition to dorsal fin | *Gasterosteus aculeatus* ☐ | |
| No spines | Go to 5 | |
| 5. Continuous dorsal fin | *Blennius pholis* ☐ | |
| Small dorsal fin | *Clupea harengus* ☐ | |

(b) Each group of living things has certain features which are important in classification. These are called distinguishing features. List four distinguishing features of fish. *(4)*

**10:18** The percentage fibre content of some common foodstuffs is shown below.

|  | % fibre |
|---|---|
| Wholemeal bread | 8.5 |
| Peanuts | 8.1 |
| Bananas | 3.4 |
| Carrots | 3.0 |
| White bread | 2.7 |
| Jacket potatoes | 2.5 |
| Cornflakes | 1.5 |

Using the graph paper on the right produce a bar chart of the information. Use a suitable scale and label for the vertical axis. *(8)*

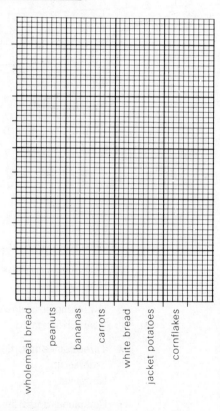

**10:19** The distributions of plant species in two land habitats were going to be compared.

    (a)  (i)  What would be the advantage of using line transects to compare the habitats? *(1)*

          (ii)  What precautions should be taken when deciding where to lay the line transects? *(1)*

(b) (i) What information could be found by using quadrats in the two habitats? *(2)*
(ii) Describe how one of these procedures could be carried out. *(2)*
(iii) How many quadrats should be thrown in each of the habitats and how should the results be used? *(1)*

(c) Why don't we just count the numbers of each plant species in the two habitats? *(2)*

**10:20** (a) A red-flowered plant was crossed with a white-flowered plant. All the $F_1$ offspring produced red flowers.
Complete the following diagram to explain these results.

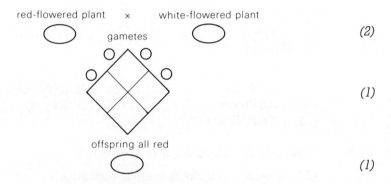

*(2)*

*(1)*

*(1)*

(b) Another red-flowered plant was crossed with a white-flowered plant. In this case half the offspring were red and half were white-flowered.
Complete the following diagram to explain these results.

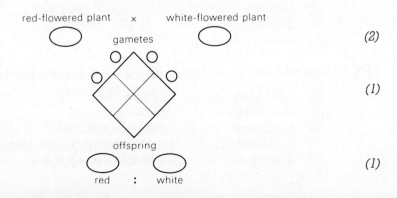

*(2)*

*(1)*

*(1)*

# Test 11

*(Answers on p. 126)*

**11:1** Bacteria need which of the following conditions in order to grow rapidly?

I    Water
II    Oxygen
III    Warmth
IV    Light

(A) I and II only
(B) I and III only
(C) I, II and III only
(D) I, II and IV only
(E) I, II, III and IV

**11:2** In the pyramid of biomass we expect to find:

(A) an equal amount of biomass at each level.
(B) that the total biomass increases up the pyramid.
(C) the greatest total biomass in the producers.
(D) the greatest total biomass in the consumers.
(E) the greatest total biomass in the tertiary consumers.

**11:3** Which of the following necessities of life are **not** re-cycled in nature?

(A) Carbon
(B) Nitrogen
(C) Sulphur
(D) Phosphorus
(E) Energy

**11:4** Which of the following pairs of items is **not** correctly matched?

(A) Intercostal muscles – move the ribs
(B) Pleural fluid – prevents friction
(C) Diaphragm – moves up and down
(D) Cartilage rings – separate thorax from abdomen
(E) Trachea – tube for the passage of air

Use the information in the table below to answer questions 11:5 and 11:6.

| Person | Body mass (kg) | Daily energy requirement (kJ) |
|---|---|---|
| (A) | 35 | 11 000 |
| (B) | 70 | 11 000 |
| (C) | 65 | 13 500 |
| (D) | 70 | 13 000 |
| (E) | 35 | 10 000 |

(A), (B), (C), (D) and (E) are five people aged 10, 11, 19, 25 and 60 (but **not** necessarily in that order).

**11:5** Which person has the highest energy requirement per unit body mass (kJ/kg)?

**11:6** Which person is **most** likely to be the 60 year old?

**11:7** Which of the following are functions of the blood?

  I   Production of red cells
  II  Production of hormones
  III Distribution of heat
  IV  Distribution of digested food

  (A) I, II and III only     (D) I and II only
  (B) I, III and IV only     (E) III and IV only
  (C) I and III only

**11:8** Which of the following things affects the diameter of the pupil?

  (A) Distance of the object   (D) Light intensity
  (B) Colour of the object     (E) External temperature
  (C) Accommodation of the eye

**11:9** What material forms the head of the long bones inside the joint?

  (A) Cartilage      (D) Muscle
  (B) Ligaments      (E) Amniotic fluid
  (C) Tendons

**11:10** Fertilization occurs in plants when the:

  (A) pollen is shed from the anthers.
  (B) pollen grains land on the stigma.
  (C) pollen grain tubes grow down the style.
  (D) pollen grain tubes enter the ovary.
  (E) pollen grain nucleus joins with the egg cell.

**11:11** Artificial selection in plant breeding can lead to:

    I    greater resistance to disease.
    II   better yields of grain.
    III  larger storage organs.
    IV  earlier flower production.

    (A)  I and II only           (D)  I, II and IV only
    (B)  II and IV only         (E)  I, II, III and IV
    (C)  I, II and III only

**11:12** Which of the following characteristics could **not** be used to distinguish between an earthworm and a centipede?

    (A)  Walking legs           (D)  Developed head
    (B)  Compound eyes        (E)  Antennae
    (C)  Segmented body

**11:13** 'A group of organisms which can interbreed' is a definition of which of the following?

    (A)  Habitat                 (D)  Environment
    (B)  Population              (E)  Ecosystem
    (C)  Community

**11:14** Assuming no other changes, population growth equals birth rate minus death rate.

If the birth rate is 5 per cent per year and the death rate is 3 per cent per year, for every 100 people at the beginning of the year how many will there be at the end of the year?

    (A)  100                     (D)  105
    (B)  102                     (E)  108
    (C)  103

**11:15** Insulin:

    (A)  leaves the pancreas in the pancreatic juice.
    (B)  is produced in too large a quantity in diabetes.
    (C)  makes glycogen change to glucose.
    (D)  is an hormonal secretion.
    (E)  is made in the adrenal glands.

**11:16** How many **sex** chromosomes are found in a human gamete?

    (A)  1                       (D)  23
    (B)  2                       (E)  46
    (C)  22

**11:17** Draw and **label** a typical plant cell and a typical animal cell to show the differences between them. *(9)*

**11:18** The number of seedlings in a field were sampled by means of several quadrats. The results are given below.

| Number of seedlings | 4 | 5 | 6 | 7 | 9 |
|---|---|---|---|---|---|
| Number of quadrats in which the above number of seedlings occurred | 2 | 2 | 4 | 3 | 1 |

(a) (i) What is the 'range' of the numbers of seedlings? *(1)*
    (ii) How many quadrats were thrown? *(1)*

(b) (i) What is meant by the 'mean' of a set of numbers? *(1)*
    (ii) Work out the mean number of seedlings in a quadrat using the information in the table. (Show your working.) *(3)*

(c) (i) If the size of the quadrat that had been used was 1m$^2$, what would be the density of the seedlings? *(1)*
    (ii) If the field sampled was 25 m$^2$, what is the estimate of the total number of seedlings in the field? *(1)*

**11:19** The figure below shows the apparatus used in a certain activity.

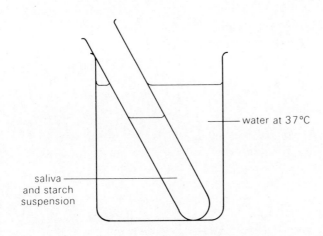

(a) What kind of reaction is taking place in the test-tube? *(1)*

(b) (i) Drops of the mixture were removed every 2 minutes and tested with iodine solution. At first the colour went blue-black, but when later samples were tested the iodine solution remained straw coloured. What can you conclude? *(2)*
(ii) How could you be sure this was the correct conclusion? *(1)*

(c) (i) Where in an animal might this reaction occur? *(1)*
(ii) What is the importance of this reaction to animals? *(1)*

(d) (i) Where in a plant might this reaction occur? *(1)*
(ii) What is the importance of this reaction to plants? *(1)*

**11:20** (a) (i) What is genetic engineering? *(1)*
(ii) Give an example. *(2)*
(iii) Describe two problems the experimenter might have when carrying out genetic engineering. *(2)*

(b) Give two reasons why it is easiest to carry out genetic engineering in bacteria. *(2)*

(c) Give two reasons why it is most difficult to carry out genetic engineering in humans. *(2)*

# Test 12

*(Answers on p. 128)*

**12:1** A cockroach nymph was 0.5 cm long. A hand lens of × 10 magnification was used to make a drawing of the nymph. The finished drawing was 2 cm long.

The magnification of the drawing was:

(A) × 0.5
(B) × 2
(C) × 4
(D) × 10
(E) × 50

**12:2** Which of the following characteristics is shown by an earthworm?

(A) Well-developed head
(B) Clitellum (saddle)
(C) Walking legs
(D) Jointed legs
(E) Antennae

Use the following diagram of an LS of a villus from the small intestine to answer questions 12:3, 12:4 and 12:5.

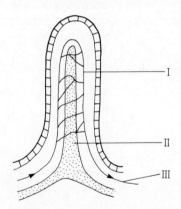

**12:3** Which of the following substances are absorbed into the part labelled I?

(A) Carbohydrates, fats and proteins
(B) Glucose, fats and amino acids
(C) Glucose, vitamins and amino acids
(D) Starch, fatty acids and amino acids
(E) Fats, vitamins and proteins

**12:4** The part labelled II is part of the:

(A) blood system.
(B) digestive system.
(C) nervous system.
(D) respiratory system.
(E) lymph system.

**12:5** The blood vessel labelled III will take blood first to the:

(A) brain.
(B) heart.
(C) lungs.
(D) liver.
(E) kidneys.

**12:6** The removal of trees is called:

(A) afforestation.
(B) deforestation.
(C) predation.
(D) monoculture.
(E) conservation.

**12:7** The table below compares photosynthesis and respiration. Which of the comparisons is **not** correct?

|     | Photosynthesis | Respiration |
| --- | --- | --- |
| (A) | Uses energy from light | Releases energy |
| (B) | Gives out oxygen | Gives out carbon dioxide |
| (C) | Goes on in the day only | Goes on in the night only |
| (D) | Occurs in green plants only | Occurs in all living cells |
| (E) | Uses water | Produces water |

**12:8** The following activity was carried out:

shoot tip covered with tinfoil cap — and lighted from the side
I

shoot tip not covered — and lighted from the side
II

The following results would be expected:

(A) Shoots I and II would bend toward the light.
(B) Shoots I and II would bend away from the light.
(C) Shoots I and II would both grow straight up.
(D) Shoot I would bend towards the light, and shoot II would grow upright.
(E) Shoot I would grow upright, and shoot II would bend towards the light.

Use the diagram of the male reproductive system to answer questions 12:9 and 12:10.

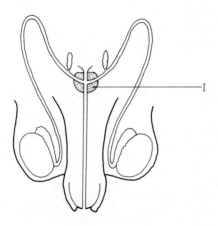

**12:9** The structure I is the:

(A) bladder.
(B) prostate gland.
(C) seminal vesicle.
(D) testis.
(E) urethra.

**12:10** The pathway along which sperm pass is:

(A) testis→sperm duct→epididymis→ureter.
(B) sperm tubules→sperm duct→seminal vesicles→ureter.
(C) testis→sperm duct→prostate gland→ureter.
(D) testis→epididymis→sperm duct→urethra.
(E) sperm tubules→prostate gland→urethra.

**12:11** In the water cycle, which of the following processes adds water to the atmosphere?

I   Respiration
II  Transpiration
III Photosynthesis
IV  Sweating

(A) I, II and III only
(B) I, III and IV only
(C) II, III and IV only
(D) I, II and IV only
(E) I, II, III and IV

**12:12** Percentage cover describes the:

(A) number of plants present.
(B) number of plants in one square metre.
(C) density of the plants.
(D) area occupied by the plants.
(E) height of the plants.

**12:13** In the nitrogen cycle which of the following might be found in root nodules?

(A) Nitrogen-fixing bacteria
(B) Denitrifying bacteria
(C) Nitrifying bacteria
(D) Decay bacteria
(E) Ammonia-converting bacteria

**12:14** Which of the following are antagonistic muscles?

I   The biceps and triceps in the arm
II  The circular and longitudinal muscles in the oesophagus
III The circular and radial muscles in the iris
IV  The erector muscles attached to the hairs in the skin

(A) I only
(B) I and II only
(C) I, II and III only
(D) I, II and IV only
(E) I, II, III and IV

**12:15** Which of the following pairs of items is **not** correctly matched?

(A) Receptacle – receives pollen
(B) Nectary – produces nectar
(C) Anther – produces pollen
(D) Ovule – produces egg cell
(E) Petals – attract insects

**12:16** The number of chromosomes in the body cells of a girl baby suffering from Down's syndrome is:

(A) 2.
(B) 22.
(C) 23.
(D) 46.
(E) 47.

**12:17** A student wanted to find out in which of **three** conditions tomatoes would ripen most quickly: on the parent plant or when removed and enclosed in a plastic bag or in a brown paper bag.

(a) How would you set up the experiment, and

(b) What problems might you have in trying to make it a fair test? *(9)*

**12:18** The diagram below shows a VS of the heart from the ventral surface.

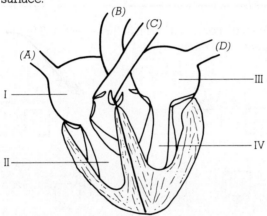

(a) Label the four chambers I–IV. *(4)*

(b) Add two arrows to the diagram to show the two blood vessels where blood **enters** the heart. *(2)*

(c) Which of the blood vessels *(A)–(D)* takes blood **to** the lungs? *(1)*

(d) Which of the blood vessels *(A)–(D)* contain oxygenated blood? *(2)*

**12:19** (a) (i) What are antibodies? *(1)*
     (ii) Where might they be formed? *(1)*

(b) (i) What are antibiotics? *(1)*
     (ii) Give an example. *(1)*

(c) When a pharmaceutical firm develops a new drug, it is first tried out on other animals before it is used on humans. Animal rights groups disagree with the testing of animals in this way.

Give one argument **for** and one argument **against** the testing of new drugs on other animals. *(4)*

**12:20** The table below shows some characteristics of the vertebrate groups.

| Group | Body temperature | Covering | Limbs |
|---|---|---|---|
| A | Warm blooded | Feathers | Wings |
| B | Cold blooded | Scales | 4 limbs |
| C | Cold blooded | Smooth skin | 4 limbs |
| D | Cold blooded | Scales | Fins |
| E | Warm blooded | Hair | 4 limbs |

(a) Use the information to fill in the letters A–E in the correct spaces in the key below.

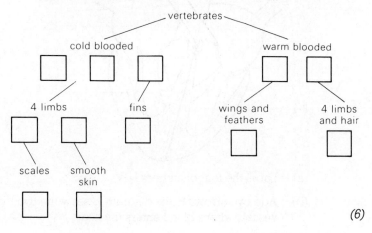

(6)

(b) What **single** characteristic in the table distinguishes the following groups?

    (i) Fish *(1)*
    (ii) Birds *(1)*

# Test 13

*(Answers on p. 130)*

**13:1** Which of the following activities occur in a green leaf?

    I    Photosynthesis
    II   Respiration
    III  Transpiration
    IV  Growth

(A) I, II and III only
(B) II, III and IV only
(C) I, II and IV only
(D) I, III and IV only
(E) I, II, III and IV only

**13:2** Which of the following comparisons between invertebrates and vertebrates is **correct**?

|     | Invertebrates | Vertebrates |
| --- | --- | --- |
| (A) | All have exoskeletons. | All have endoskeletons. |
| (B) | Adults have various numbers of appendages. | Adults all have fins or four limbs. |
| (C) | All are primary consumers. | All are secondary or tertiary consumers. |
| (D) | None have blood. | All have blood. |
| (E) | None have hearts. | All have hearts. |

**13:3** Which of the following would **not** be a correct description of bread mould?

(A) A saprophyte
(B) A fungus
(C) A lichen
(D) A decomposer
(E) A heterotroph

**13:4** 50 cm$^3$ of soil were mixed in a measuring cylinder with 50 cm$^3$ of water. The final volume of soil and water was 85 cm$^3$.
    The percentage of air in the original soil was:

(A) 10 per cent
(B) 15 per cent
(C) 30 per cent
(D) 50 per cent
(E) 85 per cent

Use the information below to answer questions 13:5 and 13:6.

Saliva was mixed with a starch suspension and divided into five parts. Each part was kept at a different temperature for 15 minutes. The amount of sugar produced at each temperature was measured.

| Temperature (°C) | 0 | 20 | 40 | 60 | 80 |
|---|---|---|---|---|---|
| Relative amount of sugar | 10 | 60 | 90 | 30 | 2 |

**13:5** At which temperature would you expect to find the most starch?

(A) 0°C  
(B) 20°C  
(C) 40°C  
(D) 60°C  
(E) 80°C  

**13:6** Which of the following conclusions can be drawn from the experiment?

(A) Enzyme activity stops at 0°C.  
(B) A very high temperature reduces enzyme activity.  
(C) Enzyme activity is unaffected by temperature.  
(D) The optimum temperature for the reaction is 20°C.  
(E) The higher the temperature the faster the rate of change of starch.  

**13:7** The skin:

I   contains sensory receptors.  
II  protects the body from micro-organisms.  
III stores excess fats.  
IV  breaks down excess amino acids.  

(A) I and III only  
(B) II and III only  
(C) I, II and III only  
(D) II, III and IV only  
(E) I, II, III and IV  

**13:8** Which of the following contraceptive methods makes use of a physical barrier to stop the sperm and egg from meeting?

(A) The sheath  
(B) The contraceptive pill  
(C) The rhythm method  
(D) Spermicides  
(E) The IUD

**13:9** In which of the following places would you expect to find meiosis occurring?

(A) An anther
(B) A root tip
(C) A shoot tip
(D) A rhizome
(E) A bud

**13:10** Which of the following statements is correct for **all** bacteria?

(A) They decay organic material.
(B) They fix nitrogen.
(C) They do not contain nuclei.
(D) They respire aerobically.
(E) They respire anaerobically.

**13:11** The number of organisms per square metre is called the:

(A) sample.
(B) community.
(C) cover.
(D) frequency.
(E) density.

**13:12** Using the relationship: rainfall (cm) = $\frac{d^2 h}{D^2}$,

calculate the rainfall in centimetres, given that:
diameter of collecting funnel ($D$) = 10 cm
diameter of measuring cylinder ($d$) = 5 cm
height of water collected ($h$) = 10 cm

The rainfall (cm) is:

(A) 0.025 cm.
(B) 0.25 cm.
(C) 0.5 cm.
(D) 2.5 cm.
(E) 5 cm.

**13:13** A protease and a 1 per cent solution of hydrochloric acid were added to a certain food in a test tube. The test tube was kept at 37°C and shaken from time to time. The food was gradually digested. What could the food have been?

(A) Sugar
(B) Cooked rice
(C) Biscuit
(D) Fish
(E) Margarine

**13:14** The production of which of the following does **not** depend upon anaerobic respiration?

(A) Humus
(B) Biogas
(C) Vinegar
(D) Bread
(E) Beer

**13:15** Which of the following is the **best** definition of homeostasis?

(A) Keeping a constant body temperature in homoiothermic animals
(B) Keeping a constant internal environment
(C) Eating healthy, wholemeal foods
(D) Controlling the water and salt levels in the blood
(E) The process by which homogenized milk is prepared

**13:16** Examples of vegetative propagation (asexual reproduction) are:

I    production of adventitious buds in leaf-of-life.
II   planting of cuttings from Busy Lizzie.
III  planting pieces of seed potatoes.
IV   production of new bulbs from buds inside an onion.

(A) II only
(B) II and III only
(C) II and IV only
(D) I, II and III only
(E) I, II, III and IV

Use the information below to answer questions 13:17 and 13:18.
In the early 1900s an experiment was carried out on rats and their diet. Rats from the same litter and of the same average mass were divided into two groups, and fed as follows.

|  | Days 0–17 | Days 18–50 |
|---|---|---|
| Group A | Purified protein, glucose, starch, fat, minerals and water only. | As before, plus 3 cm³ milk per day. |
| Group B | As for group A, plus 3 cm³ milk per day. | Purified protein, glucose, starch, fat, minerals and water only. |

The results are given below:

| | Average mass (g) | | | | | | | | | | |
|---|---|---|---|---|---|---|---|---|---|---|---|
| Days | 0 | 5 | 10 | 15 | 20 | 25 | 30 | 35 | 40 | 45 | 50 |
| Group A | 45 | 48 | 52 | 50 | 46 | 50 | 60 | 65 | 70 | 75 | 82 |
| Group B | 45 | 55 | 64 | 73 | 80 | 85 | 86 | 87 | 87 | 82 | 75 |

**13:17** (a) Using the graph paper opposite prepare a graph to show how the average masses of groups A and B varied. Add two arrows at day 18 to show where the diets were changed over. *(7)*

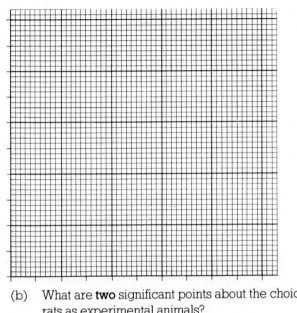

(b) What are **two** significant points about the choice of these rats as experimental animals? *(2)*

**13:18** (a) (i) Describe by reference to the graph the changes in average mass which occur in group A rats before and after day 18. *(3)*

(ii) Describe by reference to the graph the changes in average mass which occur in group B rats before and after day 18. *(3)*

(b) (i) What conclusion can be drawn from this experiment, and *(1)*

(ii) of what importance is it to humans? *(1)*

**13:19** A gardener had two red-flowered plants, A and B. He self-pollinated each one.

(a) (i) He found that plant A only gave red-flowered offspring. Explain this. *(2)*

(ii) Which genetical term would you use to describe plant A? *(1)*

(b) (i) He found that plant B gave some red- and some yellow-flowered offspring. Explain this. *(3)*

(ii) In what **proportion** (ratio) would red- and yellow-flowered plants be formed? *(1)*

(iii) If 100 plants were formed **how many** would you expect to be red-flowered? *(1)*

**13:20** You are an Agricultural Officer who has been called in to settle a dispute between a farmer and an industrial firm. The farmer is accusing the industrial firm of emptying dangerous chemicals into a river which runs through his land (see figure below).

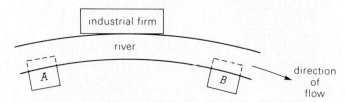

The farmer wants you to sample in areas A and B in order to settle the dispute.

(a) Why is it necessary to take samples from areas A **and** B? *(5)*

(b) After having sampled in the two areas how would you use the information you have collected to help you to settle the dispute? *(4)*

# Test 14

*(Answers on p. 133)*

**14:1** Which of the following statements is correct for **all** viruses?

- (A) They live in other organisms.
- (B) They are destroyed by antibiotics.
- (C) They are bigger than bacteria.
- (D) They respire anaerobically.
- (E) They produce spores.

**14:2** What is the meaning of the phrase 'balance of nature'?

- (A) There should be equal numbers of all kinds of organisms so that the habitat is in balance.
- (B) The interaction between organisms is such that their numbers remain fairly constant.
- (C) The biomass of producers should equal the biomass of consumers.
- (D) The number of producers should equal the number of consumers.
- (E) The decomposers should recycle the dead remains of other organisms.

Use the diagram on the right of the nitrogen cycle to answer questions 14:3 and 14:4.

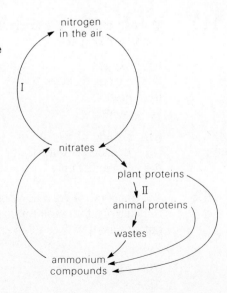

**14:3** What process is occurring at stage I in the cycle?

(A) Nitrification
(B) Denitrification
(C) Respiration
(D) Nutrition
(E) Decay

**14:4** What process is occurring at stage II in the cycle?

(A) Nitrification
(B) Denitrification
(C) Respiration
(D) Nutrition
(E) Decay

**14:5** Which of the following comparisons of typical plant and animal cells is **not** correct?

| | *Typical plant cells* | *Typical animal cells* |
|---|---|---|
| (A) | No mitochondria | Have mitochondria |
| (B) | Have large vacuoles | Have small vacuoles |
| (C) | Have chloroplasts | No chloroplasts |
| (D) | Store starch | Store glycogen |
| (E) | Have a cell wall | No cell wall |

**14:6** Respiration differs from photosynthesis because respiration:

(A) is an important life process.
(B) occurs only in animals.
(C) occurs only in plants.
(D) occurs only at night.
(E) releases energy.

**14:7** The excretory products which pass from the foetus into the mother's blood are:

I   oxygen.
II  carbon dioxide.
III urea.
IV  water.

(A) I, II and III only
(B) II, III and IV only
(C) I, II and IV only
(D) I, III and IV only
(E) I, II, III and IV

**14:8** The genotype of a red-flowered plant was Rr (with R dominant to r). If this plant was self-fertilized what would be the chance of getting plants with red flowers?

(A) 0%
(B) 25%
(C) 50%
(D) 75%
(E) 100%

The diagram on the right shows the blood supply of the liver. Use it to answer questions 14:9 and 14:10.

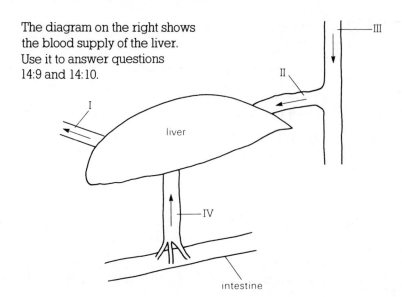

**14:9** Which blood vessels contain the most oxygen?

(A) I and II
(B) III and IV
(C) II and III
(D) I and IV
(E) II and IV

**14:10** Which blood vessels contain carbon dioxide and food?

(A) I and II
(B) II and III
(C) III and IV
(D) I and IV
(E) II and IV

**14:11** The tongue is sensitive to:

I sweet.
II sour.
III touch.
IV heat.

(A) I and II only
(B) III and IV only
(C) I, II and III only
(D) I, III and IV only
(E) I, II, III and IV

**14:12** Part of the biotic environment of an insect would be:

(A) the air which it breathes.
(B) the place where it lives.
(C) the excreta that it produces.
(D) the flowers which it pollinates.
(E) the water that it needs to conserve.

**14:13** Which of the following statements are correct descriptions of compost?

  I   It can be made by the farmer.
  II  It releases mineral salts slowly.
  III It improves the soil texture.
  IV  It reduces water loss.

  (A) I, II and III only        (D) II, III and IV only
  (B) I, II and IV only         (E) I, II, III and IV
  (C) I, III and IV only

**14:14** A deficiency in which of the following elements is likely to cause the greatest reduction of growth in plants?

  (A) Nitrogen          (D) Potassium
  (B) Magnesium         (E) Calcium
  (C) Iron

**14:15** Which of the following comparisons between arteries and veins is **not** correct?

|     | Arteries | Veins |
| --- | --- | --- |
| (A) | Thick wall | Thin wall |
| (B) | Always take blood away from heart | Always take blood towards heart |
| (C) | Always transport oxygenated blood | Always transport deoxygenated blood |
| (D) | Valves absent | Valves present |
| (E) | Narrow central hole | Wide central hole |

**14:16** A mutation occurs in the petals of a certain flowering plant. This mutation:

  (A) can be passed on if it occurs before pollination.
  (B) can be passed on if it occurs in the maternal plant.
  (C) can be passed on if it occurs before fertilization.
  (D) can be passed on if the plant is self-fertilized.
  (E) cannot be passed on.

**14:17** The following diagram shows an experiment set up to test the conditions necessary for photosynthesis.

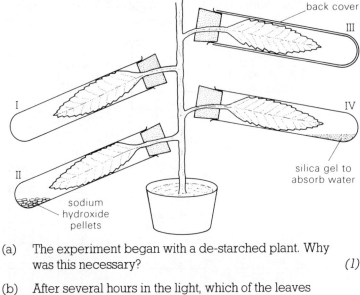

(a) The experiment began with a de-starched plant. Why was this necessary? (1)

(b) After several hours in the light, which of the leaves would show the presence of starch? (2)

(c) What is the purpose of each of the arrangements I, II, III and IV? Write a sentence about each one. (4)

(d) (i) What is a control? (1)
    (ii) Why is it necessary to have a control? (1)

**14:18** The diagram below shows a vertical section of some of the structures in the abdomen of a female mammal.

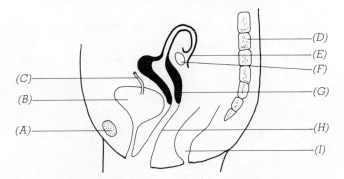

(a) Identify and name the four parts of the **female reproductive system**. (4)

(b) Write a sentence about **each** of the four parts to describe their functions. (4)

**14:19** The diagram below shows the distribution of three plant species, X, Y and Z.

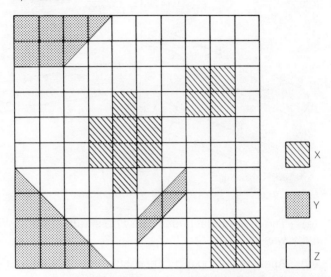

(a) Work out the frequency of each species, X, Y and Z. *(3)*

(b) Work out the percentage cover of each species, X, Y and Z. *(3)*

(c) (i) What is the density of plants? *(1)*
(ii) If each small square covered by species X represents two plants, what is the density of species X in the square metre shown? *(1)*

**14:20** Imagine that you have been given five animals of various sizes to classify. Describe the procedure you would use as you prepare a key to separate the animals. *(9)*

# Test 15

*(Answers on p. 135)*

**15:1** Which of the following animals is cold blooded (poikilothermic) but shows adaptations in its behaviour which help it to maintain a more constant body temperature?

(A) Fish
(B) Toad
(C) Lizard
(D) Bird
(E) Mouse

**15:2** Perch, which are bred in fish farms as a source of protein for humans, eat insect larvae, which feed on microscopic algae in the water.

The organisms mentioned above are:

I  perch.
II  humans.
III  insect larvae.
IV  microscopic algae.

Which of the following is a correct food chain with the producer listed first?

(A) I, II, III, IV
(B) IV, III, II, I
(C) I, III, IV, II
(D) IV, III, I, II
(E) III, IV, I, II

**15:3** 'Overgrazing' means:

(A) the same as 'overfishing'.
(B) having two kinds of animals grazing in the same field.
(C) using a field for grazing instead of growing crops.
(D) using a field for grazing after having grown a crop.
(E) having too many cattle grazing in a field.

**15:4** Which of the following contraceptive methods is most effective for preventing conception?

(A) Vasectomy
(B) Using an intra-uterine device
(C) Using a diaphragm with spermicides
(D) Using the contraceptive pill
(E) Avoiding 'unsafe' days

**15:5** If the magnifying power of the eyepiece lens is × 10, and that of the objective lens is × 10, what is the total magnification of the microscope?

(A)  × 10
(B)  × 20
(C)  × 30
(D)  × 100
(E)  × 1000

The diagram below shows a section of part of the lower epidermis of a leaf. Use the information to answer questions 15:6 and 15:7.

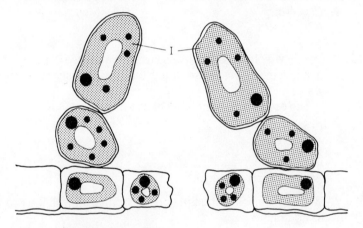

**15:6** The cells labelled I on the diagram are:

(A) epidermal cells.
(B) guard cells.
(C) spongy mesophyll cells.
(D) palisade cells.
(E) stomata.

**15:7** In the daytime:

(A) the guard cells and epidermal cells carry out photosynthesis.
(B) the guard cells produce sugars and become turgid, so opening the stomata.
(C) the guard cells produce sugars and become plasmolysed, so opening the stomata.
(D) the stomata are closed to stop the escape of water needed for photosynthesis.
(E) the stomata are closed to stop the escape of carbon dioxide needed for photosynthesis.

**15:8** Arteries have thicker and more elastic walls than veins. This is because the blood in arteries:

(A) contains more food.
(B) contains more oxygen.
(C) is under more pressure.
(D) has to go to all parts of the body.
(E) is thicker.

**15:9** A girl was reading her book. She then looked out of the window to focus her eyes on a tree. The lenses in her eyes became:

(A) thinner.
(B) fatter.
(C) harder.
(D) softer.
(E) more opaque.

**15:10** Human reproduction to produce a baby involves:

I   mitosis.
II  meiosis.
III asexual reproduction.
IV  sexual reproduction.

(A) I and II only
(B) I, II and III only
(C) I, II and IV only
(D) II and IV only
(E) I, II, III and IV

**15:11** An organism was crawling around on the ground. It was soft and elongated. Its body was divided into about thirteen segments. It had no legs.
    It could have been a:

I   tapeworm.
II  small earthworm.
III fly maggot.
IV  caterpillar

(A) I and II only
(B) II and III only
(C) III and IV only
(D) I and IV only
(E) II and IV only

**15:12** A red-flowered homozygous plant was crossed with a white-flowered homozygous plant. If the red allele is dominant to the white allele, the offspring will be:

(A) all red-flowered.
(B) all white-flowered.
(C) all pink-flowered.
(D) half red-flowered and half white-flowered.
(E) half red-flowered and half pink-flowered.

**15:13** Which of the following, when dissolved in water, is the **main** cause of 'acid rain'?

(A) Carbon monoxide
(B) Carbon dioxide
(C) Sulphur dioxide
(D) Soot
(E) DDT

**15:14** The apparatus shown below was used to test inhaled and exhaled air.

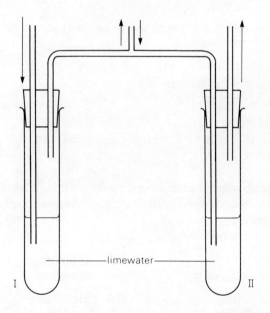

Test-tube I is important because it:

(A) tests the exhaled air.
(B) removes any carbon dioxide from the air.
(C) filters out germs.
(D) shows the presence of carbon dioxide.
(E) acts as a control.

Use the graph on the right, which shows the growth of a human foetus before birth, to answer questions 15:15 and 15:16.

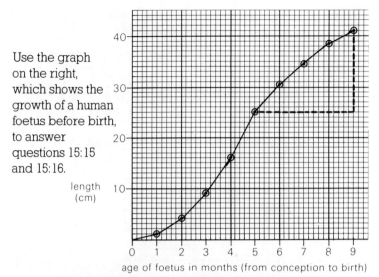

age of foetus in months (from conception to birth)

**15:15** What is the rate of growth (in cm/month) of the foetus in the sixth and ninth months (as shown by the dotted lines)?

(A) 4 cm/month  (D) 7 cm/month
(B) 5 cm/month  (E) 8 cm/month
(C) 6 cm/month

**15:16** In which month is the rate of growth **most** rapid?

(A) First month  (D) Seventh month
(B) Third month  (E) Ninth month
(C) Fifth month

**15:17** The table below shows the percentages of the main substances in the body.

| Main substances in the body | Percentage |
|---|---|
| Water | 65 |
| Protein | 20 |
| Fat | 10 |
| Carbohydrate | 5 |

A student is going to make a pie chart of this information. Work out the angles needed to represent each of the sectors.
    Show your working.

(i)   Fat                                     (2)
(ii)  Carbohydrate               (2)
(iii) Protein                         (2)
(iv) Water                           (3)

**15:18** The table below shows the composition of a full-cream milk and a semi-skimmed milk of a certain brand per 100 cm³.

|  | Fat (g) | Protein (g) | Carbohydrate (g) |
|---|---|---|---|
| Full-cream milk (Fc) | 3.9 | 3.4 | 4.8 |
| Semi-skimmed milk (Ss) | 1.0 | 3.8 | 5.5 |

(a) Using the graph paper below produce a bar chart comparing the two kinds of milk. *(6)*

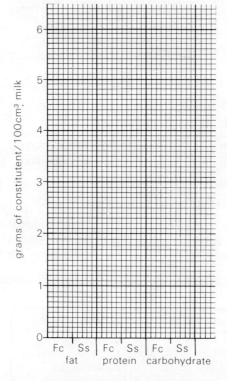

(b) One of the milk samples provides 190 kJ of energy per 100 cm³, while the other provides 281 kJ per 100 cm³. Which one do you think provides 190 kJ/100cm³? Explain your answer. *(2)*

**15:19** In the 1950s an experiment was done with light- and dark-winged peppered moths. Moths of each kind were released in an industrialized city and in a country area. They were marked with small spots of paint and some of them were later recaptured.

How long had they been doing this? Brontë wondered as she felt the minutes tick by. How long could they continue doing it before they would have to concede Mr Finlay was not going to come back?

'I don't give up easily, Brontë,' Eli muttered as though he'd read her mind. 'Paddles again.'

Obediently, Brontë handed them to him and, as Eli placed them on either side of Mr Finlay's chest, and the electric shock ran through him, the heart monitor's erratic recordings suddenly changed. Miraculously, they had a pattern. It was much too slow, much too deep, but at least it was regular.

'Pulse?' Eli demanded.

'Weak and slow, but there,' Brontë replied.

'Okay, we go,' Eli declared, and she nodded.

They might have got Mr Finlay back, but his heart could return to VF at any minute. He needed A and E, but first they would have to negotiate three flights of stairs, and Mr Finlay was a big man. A very big man.

'How strong is your back?' Brontë asked as Eli stood, and Mrs Finlay appeared carrying an armful of blankets, and two flasks.

'I was going to ask you the same thing,' Eli

replied wryly. 'Thank heavens for the carry-chair. Trying to get him down those stairs on a stretcher would have been a nightmare.'

It wasn't much easier with the carry-chair, Brontë decided. By the time they'd reached the ground floor she was breathless, and sweaty, and her arms and shoulders felt as though they had been pulled out of their sockets.

'Blue this one in, Brontë,' Eli ordered, as he climbed in beside Mr Finlay and his wife at the back of the ambulance. 'Up the siren to max, and step on it!'

She did, but it wasn't an easy ride. According to their sat nav, the quickest way to the hospital was via some of Edinburgh's back roads, but those streets also contained a hazard she hadn't been prepared for.

'I'm sorry—so sorry,' she shouted over her shoulder as she heard Eli swear when the ambulance lurched jerkily to one side yet again. 'If I could avoid these wretched speed bumps, I would, but I can't.'

'And if I had my way I'd strap every Edinburgh councillor to a trolley, then put him in an ambulance and drive them over these damn things for ten minutes, and you can bet your life they'd have them all dug up before I could say Elijah Munroe!' he exclaimed.

Brontë could not help but agree with him, and nor could she restrain her sigh of relief when she saw the lights of the Pentland Infirmary shining through the dark in front of them. Their journey might only have taken ten minutes, but it had seemed like a lifetime.

'I hope Mr Finlay makes it,' she observed after the nurses and doctors had taken over from them, and whisked Mr Finlay and his wife away.

'We did our best,' Eli replied with a tired smile, 'and it's all we can do. Which reminds me,' he added. 'When we were coming down the stairs, I noticed—'

'I didn't always bend my knees,' she finished for him. 'Yes, I know. I'm out of practice lifting people.'

'It's not that,' he said. 'It's your boots.'

'I bought a new pair,' she protested. 'I went to Harper & Stolins this afternoon...' She glanced at her watch. 'Actually, yesterday afternoon now, and I bought a pair.'

'Those are not Safari boots,' he said firmly. 'You may well have bought them from Harper & Stolins but they are not Safari.'

'The Safari ones were too heavy,' she replied. 'The Wayfarer ones were lighter, more fashionable—'

'There's a reason for them being heavier,' he interrupted with great and obvious patience. 'It's to ensure you keep your toes.'

'Okay, all right.' She grimaced. 'So I bought the wrong boots, but must you always be so…so…'

'Right about everything?' he suggested with a smile.

'Smug,' she countered. 'Smug was the word I was searching for. I'll go back this afternoon and buy a pair of Safaris.'

'What time this afternoon?'

She had to think. Working nights was really throwing her awareness of time completely.

'I'll probably be sleeping until about two,' she replied, 'then I need to do some washing, and buy some food…. As tonight's late-night shopping, I'll probably go about eight o'clock.'

'Then I'll meet you outside Harper & Stolins at eight o'clock to make sure you get the right boots this time. And before you argue with me,' he added as she opened her mouth to do just that, 'I'm damned if I am going to have to live for the rest of my life with the knowledge that, because I didn't supervise you, you lost all your toes.'

'But—'

'Eight o'clock, Brontë.'

She gave in. 'Okay, I'll meet you outside the shop. Happy now?'

He was until they got back into the ambulance, and then she heard him groan.

'What's up?' she asked.

'That last call-out. It was a failure.'

'Mr Finlay?' She faltered. 'You mean… He's died?'

'I don't know,' Eli replied. 'It's one of the downsides of being a paramedic. Unless we specifically make enquiries, we never get to find out what's happened to the people we pick up.'

'Then what are you talking about?' she protested. 'We arrived at Holyrood Gate to find Mr Finlay wasn't breathing, you got his heart beating again, we've delivered him to A and E, so you've given him the best possible chance of survival. No way is that a failure.'

'It is if you look at the dashboard,' Eli declared. 'I hit the timer when we got the call-out, and hit it again when we arrived in Holyrood Gate, and

I've only just checked it. Look at the reading, Brontë.'

She did, but it didn't help.

'I'm sorry, but I'm still not with you,' she said.

'Brontë, it was a cat-A, high-priority call. If you remember, an ambulance crew should arrive at a cat-A call-out in eight minutes. It doesn't matter what happens to the patient *after* we get there, just so long as we get there in eight minutes, and we took nine minutes.'

She gazed at him in disbelief. 'Then, you're saying…'

'If we had arrived at Holyrood Gate in eight minutes to discover Mr Finlay had been dead for two days, it would be counted as a success. If we arrived, as we did, in nine minutes, providing him with life-saving treatment, it's a failure.'

'But that's…' She thought about it, but thinking didn't make it any better. *'Nuts.'*

'Yup,' he agreed, wearily rubbing his hands over his face, 'and it's not just nuts. It also costs the station dearly because the more eight-minute targets we miss, the less money the government will give us to buy new vehicles, and employ more staff. Mr Finlay's call-out was a financial disaster for ED7.'

All the elation Brontë had felt in getting Mr Finlay safely to the Pentland drained away in an instant. She knew she couldn't possibly have got to Holyrood Gate any faster. No one could unless they'd had wings, and as she stared unhappily at Eli, she realised he suddenly looked every one of his thirty-eight years.

'How do you stand this?' she asked. 'The stupid rules, the petty restrictions?'

'Because…' He drew in a deep breath, then shook his head awkwardly. 'Hell, but this is going to sound so pretentious. I might hate the bureaucracy, the reducing of patients to numbers instead of people with feelings, hopes and dreams, but… Every once in a while I make a difference. Every once in a while my being there can pull someone through who wouldn't otherwise survive, and on those occasions…'

'It's the best job in the world,' she said, and he smiled, a weary, sad smile.

'That's about it.'

It was how she'd used to feel in A and E, she remembered. The rush of adrenaline when they'd been able to resuscitate someone, the buzz in the department when everything went well. Yes, there were downsides. The consultants who could be rude and overbearing, the times when they couldn't save someone, but she missed it, she missed it so much.

'Coffee,' she said firmly. 'I think we both need coffee, and I want the biggest, stickiest, sugar-covered doughnut Tony's can supply.'

'Not a hamburger?' Eli said, his blue eyes crinkling, and she knew the effort it had taken him to make that small joke, and gave him a mock hard stare.

'A doughnut, Mr Munroe.'

'Then hit the road, O'Brian,' he replied, 'and don't spare the horses.'

She didn't. She reached the café in record time but, when Eli had gone to get their food and drinks, she stared morosely out of the ambulance.

How in the world had she ever believed she could do this job? It had been a mistake even to have applied for it. She should have looked into the details more fully, paid more attention to what her duties would be, but she had been so desperate for a job, any job.

*And you wanted something which would keep*

*you connected, however tenuously, to nursing,* her mind whispered, and she sighed.

Not once since she'd started working at the station had she ever taken out her notebook because it was the patients who interested her, they always had. Nursing was where her heart lay and, though Eli seemed to think she could retrain to be a paramedic, unless she could get over her fear of crowds it just wasn't an option, and that left her…

Nowhere, she thought with an even deeper sigh. Absolutely nowhere.

'One-a cappuccino, one-a doughnut,' Eli announced with a flourish as he opened the cab door.

'You're a lifesaver,' she replied, forcing a smile to her lips, but he wasn't fooled for a second.

'Let it go, Brontë,' he said gently. 'Take a tip from one who knows. The petty restrictions, the rules and regulations, let them go and concentrate on the fact we got Mr Finlay to hospital because, if you don't, it eats away at you, makes you cynical, bitter.'

'I know,' she murmured, 'but…' She shook her head. 'I don't think I'm cut out for this job.'

'I don't think you are either,' he agreed, and she glowered at him.

'Couldn't you at least *pretend* to think I might be?' she exclaimed. 'Sheesh, but you really know how to dent a woman's self-esteem, don't you?'

'Would you rather I'd lied?' he asked, and she bit her lip.

'No, but you could have glossed it up a bit,' she pointed out. 'You could have said, "Well, Brontë, I think you're really good at this job, but I just know you'd be even better doing something else."'

He grinned. 'Yeah, but the big question is, would you have bought it if I'd said that?'

She wouldn't, but it was singularly depressing to know this was the second job she'd failed at within the space of a year.

'If you decide this isn't right for you,' he continued as she took a morose bite out of her doughnut, 'how about retraining to become a paediatrics nurse, or something in surgical?'

'I could, except…' She grimaced slightly. 'I don't mind kids, but nursing them all the time… And as for surgical… There wouldn't be the same buzz I used to get from A and E.'

'Brontë—'

'Anyway, it's not your problem, is it?' she said brightly. 'So let's talk about something else.'

She was right, he realised, it wasn't his problem so why did he feel so concerned? What she did, where she went after the end of this week, was surely her own business, and it shouldn't matter to him what decision she made, and yet, to his acute annoyance, he discovered it did.

*Protective.* Hell but he had that feeling again, and it didn't make any sense. He had never once felt protective of any his ex-girlfriends, so why in the world was he feeling it now about a woman he barely knew?

Because she's had such a rotten time this past year, he told himself. She's been stabbed, she suffers from panic attacks, and she's going to be out of a job because she sure as heck can't do an assessor's one. Only a complete louse wouldn't feel sorry for her.

*Except feeling sorry isn't the same as feeling protective.*

'Why are you glowering at me?'

'Glowering?' he repeated blankly, and saw Brontë shake her head.

'You're glowering at me, like I've rained on

your parade or something, so what have I done wrong now?'

No way was he going to tell her his thoughts, and so he said the first thing that came into his head.

'Why Brontë?'

'Why Brontë, what?' she asked, taking a sip of her coffee.

'No, I meant how did you get your Christian name?'

'Oh, that's easy,' she replied. 'My parents both lecture in, and are fanatical fans of, English nineteenth-century literature, so I got stuck with Brontë. Actually, in the greater scheme of things, I was lucky. My big brother got landed with Byron, and my little sister with Rossetti.'

'And are your brother and sister medics like you?'

'Oh, heavens, no.' She laughed. 'Byron is an investment banker, and Rossetti's a criminal lawyer.'

His eyebrows rose. 'Serious high-flyers.'

'Yup, I'm the dummkopf of the family,' she replied, and saw a flash of unexpected irritation cross his face.

'Why do you think you're the dummkopf?' he demanded.

'Because I am in comparison to them,' she replied. 'Look, it's no big deal,' she continued, as he opened his mouth, clearly intending to interrupt, 'we can't all be high-flyers and I don't have a problem with it. Do you have any brothers or sisters?'

'No.'

An oddly shuttered look had suddenly appeared on his face. A look which suggested she had inadvertently strayed into an area he considered very much off limits, and she wished she hadn't asked, except it was hardly a controversial question, and yet it appeared it was.

'Good coffee,' she said awkwardly. 'Good doughnut, too.'

'Maybe I'll tempt you into trying one of Tony's hamburgers one night,' he said, clearly deeply relieved to be talking about something else. 'Better yet, maybe you'll tell me about the blond who did such a number on you that you've given up dating permanently.'

She rolled her eyes.

'Don't you *ever* give up?' she exclaimed.

'Nope.'

Which would teach her the folly of embellishing a lie, she thought ruefully. There wasn't a blond, never had been unless she counted the fair-haired medical student she'd dated for a few weeks when she was training to be a nurse, but how to explain what she didn't fully understand herself. That as far as men were concerned she seemed to have an invisible sticker on her forehead saying, *This one's a mug*. Eli knowing she was a fruit cake when it came to encountering drunken gangs was one thing, her admitting to him she was also the world's biggest all-time loser in the game of love was something else entirely.

'Next question, please,' she said firmly.

'Quit stalling, O'Brian,' he pressed. 'Tell me.'

She wouldn't have told him anything, but he was smiling that smile at her again. The 'tell me all your troubles' smile, the 'you can trust me' smile, and she gazed heavenwards with frustration.

'Look, if I tell you, will you quit harassing me?' she demanded, and when he nodded, she took a deep breath. 'It's got nothing to do with any blond. I just have such lousy taste in men I thought, Why keep on getting hurt, why not just concede defeat, and give up on dating completely.'

A frown appeared in his blue eyes. 'You mean, you attract men who hurt you physically?'

'No, I don't mean that,' she said irritably.

'Then you attract psychos, weirdos?'

'You mean like the loony tune I'm currently sitting next to?' she said in exasperation. 'No, I don't mean that either. It's just…every man I've ever dated… It always starts off okay, and then…' She shrugged. 'I guess I don't see the warning signs quickly enough that the men I get involved with aren't right for me.'

'Why?'

'I don't *know*. Look, you asked me,' she continued quickly as Eli opened his mouth clearly intending to push it, 'and I've answered your question. *End of discussion.*'

Not for Eli it wasn't, she realised as he waved the remains of his hamburger pointedly at her.

'Okay, as I see it,' he declared, 'you're making two fundamental mistakes here. First, you're choosing the wrong men to date.'

'Well, duh,' she replied. 'I wonder why I didn't think of that. *Of course* I'm choosing the wrong men to date, but I don't deliberately set out to fall in love with men who will break my heart. I just seem to end up with them.'

'And that's your second mistake,' Eli observed. 'You're looking for *lurve*, for the happy-ever-after ending, and there's no such thing.'

She blinked. 'You don't believe there's any such thing as love?'

'Of course there isn't,' he replied, taking a large gulp of his coffee. 'All that hearts and flowers stuff, the mushy sentimental ballads and films… It all boils down to sex.'

'But—'

'Brontë, the quicker you wake up and face reality, the less chance you'll have of being hurt,' he insisted. 'Forget about love, forget about happy-ever-afters. Relationships between men and women all come down to one thing. What the sex is like. If the sex is mediocre, you cut your losses. If the sex is okay, you stick around for a bit because okay sex is better than no sex at all, and if the sex is phenomenal you enjoy it while you can, and then move on. It's what grown-up, realistic men and women do.'

'No, you're wrong, so wrong,' she argued back. 'There *is* love in this world. Yes, there's a lot of suffering, a lot of pain, but I truly believe there are people out there caring for one another, loving one another. If I didn't believe that—if I just accepted, as you seem to, that everyone just looks after themselves—what kind of world would we have?'

Eli shook his head impatiently.

'You're confusing two completely different issues here,' he declared. 'I'm not saying you walk on by if someone's in trouble, or ignore another human being's suffering. There are a lot of things I care passionately about. Injustice, inequality, bigotry—'

'But not love?' she interrupted.

'No, not love,' he said firmly, and she sighed in defeat.

'Could any two people possibly be more incompatible?' she observed. 'More poles apart in what they think, believe?'

'Doubt it,' he replied.

There didn't seem anything left to say, and, as Brontë stared down at the remains of her doughnut, it somehow didn't seem nearly as appealing as it had earlier.

'I think I've had enough of this,' she said, putting what was left of her doughnut back into its paper bag.

'I'm actually not very hungry tonight either,'

Eli murmured, gazing down at his hamburger without enthusiasm.

'Back to the station, to wait for a call-out?' she suggested.

'Best thing,' he agreed. 'Except you'd better get rid of that icing sugar on your cheek first. No, the other cheek,' he continued impatiently as she put her hand up to her left cheek. 'Here, let me.'

She didn't get the chance to tell him she wasn't an idiot, that she could manage perfectly well on her own, thank you very much. He had already begun brushing the icing sugar from her cheek. Brushing it matter-of-factly at first, and then the pressure of his fingers suddenly changed. His touch became feather-light, almost caressing, and she forgot to breathe, forgot to do anything, but stare back at him. He hadn't moved closer to her, she could swear he hadn't, and yet he seemed so much nearer, the cabin so much smaller, and, when his fingers slid slowly down her cheek, and cupped her chin, she swallowed hard.

'Eli….'

She couldn't say anything else, and he didn't say anything at all. He just held her chin in his hand, his blue eyes fixed on her, so dark, so very dark in the light from the street lamp, and she could hear his breathing, could hear her own erratic breath in the silence, could feel a slow spiralling heat growing deep within her, and slowly she reached up and covered his hand with her own.

His fingers were warm, much warmer than hers, and she shivered involuntarily, saw his lips part, knew her own had, too, and she felt herself leaning towards him, saw he was leaning towards her in return, and then, suddenly, without warning, he wasn't holding her chin any more, and she was looking shakily out of the window, not sure whether it had been her, or him, who had moved first, or if they had moved in unison, but it was definitely Eli who spoke first, and his voice, when he did, sounded strained, husky.

'We'd better get back to the station.'

She nodded, or at least she thought she nodded, and with her heart still jumping erratically in her chest, she switched on the ignition and hoped to heaven he couldn't see just how much her hands were trembling as she drove away.

# CHAPTER FOUR

*Thursday, 7:55 p.m.*

SHE should have arranged to meet Eli earlier, Brontë thought as she stamped her feet, and blew on her fingers to try to restore some heat to them, as an icy wind swirled around Cockburn Street. Better yet, she should have told him at the end of their shift this morning that she'd suddenly remembered something absolutely vital she had to do, and couldn't meet him at all.

Not that they'd exactly been talking when they'd finished work, she remembered ruefully. He'd seemed as anxious to get away as she had, which meant this shopping trip was going to be awkward, and uncomfortable, and she should have stayed home.

'Oh, stop it, Brontë,' she muttered to herself. 'It's not like this is a date. You've both agreed you're completely incompatible, and him coming on this shopping trip is more a colleague-to-colleague advisory type thing.'

*Which doesn't explain why you've washed your hair, and put on your favourite knee-length brown leather boots, and equally favourite racing green coat with the faux-fur-trimmed hood.*

Well, my hair needed a wash, she defensively told her reflection in the large window of Harper & Stolins, and as for my clothes… Only an idiot wouldn't have wrapped up warmly when it definitely felt like snow.

*Yeah, right, Brontë,* a little voice laughed as she tried in vain to smooth down her fringe which was most definitely sticking up. *So it's got nothing to do with what happened earlier this morning, then?*

Oh, my word, but that had been something else, Brontë thought, feeling a flutter of heat in her stomach as she remembered. When he'd touched her cheek, when their eyes had met, and time had seemed to disappear… She'd been so certain he was going to kiss her, had so wanted him to kiss her, and she closed her eyes, to relive the memory, and let out a tiny yelp when a hand clasped her shoulder.

'Will you stop *doing* that?' She gasped, clutching at her chest as she whirled round to see Eli standing behind her.

'I thought it was just dark cemeteries which spooked you,' he protested, 'not well-lit streets in Edinburgh.'

Everything about you spooks me, she wanted to reply, but she didn't.

'Yeah, well, wear shoes with heavier soles next time,' she said instead.

In fact, change absolutely everything you're wearing, she thought as her gaze took in his appearance. Eli dressed in his paramedic cargo trousers, and high-visibility jacket, might be enough to set most feminine hearts aflutter, but Eli wearing a pair of hip-hugging denims, a blue-and-white open-necked shirt and an old black leather jacket was guaranteed to give every woman a cardiac arrest. A fact that was all too obvious from the number of women who were giving him second glances as they walked by.

'So,' he said, 'are you ready for the big boots expedition?'

Was it her imagination or did he seem just a little bit uncomfortable, a little unsure, almost as though he wished he was anywhere but here?

'Look, you don't have to do this,' she said

quickly. 'I promise faithfully I'll buy the correct boots this time, so you don't need to babysit me.'

'Hey, it's no problem,' he insisted. 'And I have to safeguard your toes, remember?'

She laughed, and he did, too, and if his laughter sounded slightly strained to her ears, she decided it was probably just because he was as cold as she was.

'Okay,' she said, 'let's get on with it, shall we?'

Getting on with it, however, looked as though it might take considerably longer than she'd anticipated. They were certainly whisked instantly through to the seating area when they entered Harper & Stolins, and there were assistants aplenty, but unfortunately each and every one of them appeared to want to talk to Eli. Of course, that was probably because all of them were women, Brontë thought wryly, but feeling like a spare parcel abandoned at a sorting office was not how she had envisaged spending her boot-buying expedition.

'Any chance of some service here?' she said eventually to one of the girls who was hovering on the outskirts of Eli's fan club.

'Sorry,' the girl declared, looking as though she actually meant it, 'but Eli's a very popular customer.'

'So I see.' Brontë sighed. 'I'm looking for a pair of Safari boots, size five.'

The girl was back within minutes, clutching a pair, and it was only when Brontë put them on she realised the folly of wearing a skirt when you were trying on safety boots.

'What's wrong?' Eli asked as he joined her beside the mirror, and saw her rueful expression.

'Not exactly flattering, are they?' she replied, looking down at her feet. 'In fact, I look like I'm auditioning for clown of the year.'

'They're not supposed to be flattering,' Eli observed. 'They're supposed to keep your toes in one piece.'

'I know, but…' She sighed as she lifted one foot. 'I should have worn trousers.'

'Not for me you shouldn't. It's nice to see you have legs.'

And he was staring at them. Staring at them in a quite blatant way, and she suddenly felt completely exposed, which was crazy because she walked around Edinburgh in skirts all the time, and had never once felt vulnerable.

'Of course I have legs,' she replied in a rush, horribly aware her cheeks were darkening. 'Everyone has legs. Unless they've lost them due to some accident, or were born that way, of course, in which case they won't but, generally, normally, most people have legs.'

And I'm babbling, she thought, babbling like an idiot, but I wish I'd worn trousers because he's still staring at my legs and my knees are too chunky, and my calves aren't exactly model-girl slim, and Eli likes girls with impossibly long legs, not that I give a damn about that, but…

'Everyone might have legs,' he said, 'but not everyone has great ones like you.'

He thought she had good legs. No, correct that. He thought she had *great* legs, and he was still looking at them, making her feel even more self-conscious.

'Do we have a sale?' the assistant asked, glancing from Brontë to Eli, then back again.

'I think so, yes,' Brontë replied. 'I mean they fit, and they're Safaris, so…'

She wished Eli would say something, anything, and what was worse was two of the assistants were nudging each other, and giggling, and she didn't know why they were giggling.

'These boots are the right ones, aren't they, Eli?' she continued pointedly. 'The ones you wanted me to buy?'

He blinked, then nodded. 'Absolutely. Definitely.'

'Well, that's my footwear sorted out for tonight,' she said far too brightly as she sat down, feeling considerably flustered. 'Problem solved, mission accomplished.'

'So, what now?' he asked as she slipped into her knee-length boots, and headed for the till, clutching the Safaris.

'Now?' she repeated. 'Well, I guess once I've paid for the boots I go home, and you go and do whatever you were planning on doing before we clock on in a couple of hours.'

'You mean I don't even get to share a celebratory dinner with you?' he protested after she'd settled up with the cashier, and he followed her out onto the street. 'I've come all the way here from my warm and cosy flat in Lauriston Place to give you the benefit of my not-inconsiderable advice, and now it's snowing, and I don't even get something to eat?'

He wanted to prolong this nondate? He wanted to go somewhere else with her? She would have thought he would have been anxious to get away, to do something else, see someone else, and yet he clearly wasn't.

*Go home, Brontë. Thank him for his kindness, and go home where it's safe.*

Except he was right about the snow. Little flakes were beginning to whirl about in the wind, and settle on the pavement, and he hadn't needed to come and help her, even if she hadn't really needed any help. All she'd needed was to concede defeat, forget about fashion and style, and accept she had to wear a pair of boots which made her feet look like a penguin's.

'Dinner would be nice,' she admitted. 'But I live on the south side of Edinburgh so I wouldn't have a clue where to go round here.'

'The Black Bull in the High Street,' he announced. 'It's my favourite restaurant. They might not make coffee quite as good as Tony's, but they do make a mean Hungarian stroganoff and poached salmon to die for.'

And he was obviously as equally well known in the Black Bull as he was in Harper & Stolins, Brontë thought drily, when a beaming waitress ushered them towards a table by the roaring log fire, and an equally attentive waitress quickly took their orders. A stroganoff for Eli, and poached salmon for her.

'Do you know every single woman in Edinburgh?' Brontë asked as she shrugged off her coat, and held her hands out to the fire. 'And I'm emphasising the word *single* here.'

'Can I help it if I'm a popular guy?' Eli answered, and, despite herself, Brontë laughed.

'It's nice here,' she said, her eyes taking in the old oak panelling, the wheelback chairs, chintz curtains, and pictures of Old Edinburgh. 'Very cosy, and atmospheric. Though I'm surprised you like it. I'd have pegged you as more a minimalist, modern sort of a guy.'

'Now, what did I tell you about the dangers of taking me at face value?' he teased and, when she laughed again, he nodded approvingly. 'You don't do that often enough. Laugh, I mean. I wonder why that is?'

'Maybe I just take things more seriously than you do,' she replied. 'Or, maybe this past year has forced me to be more serious.'

To her surprise, Eli reached out, and covered one of her hands with his.

'Try not to remember it,' he said gently. 'Try not to look back, but look forward instead.'

Which was easy for him to say, Brontë thought wistfully as she stared down at his strong, capable fingers. He knew exactly what his future was, whereas she didn't even know what she would be doing next week.

'By the way, I think I've figured out why you're having such problems with men,' Eli continued, releasing her hand as the waitress placed their orders in front of them.

'I thought we'd already worked that one out,' she protested, trying not to mind that his hand was no longer covering hers. 'It's because I'm a lousy picker, and I believe in love which, according to you, doesn't exist.'

'Well, those are certainly issues, but I think they're part of a much bigger problem. How old is your brother?'

'Byron?' she said, bewildered by his unexpected change in conversation. 'He's thirty-six.'

'And your sister, Rossetti?'

'She's twenty-eight, but...' She frowned at him. 'Sorry, but I thought you were going to explain why I have such bad luck with men, but now you're talking about my family, so is there a reason behind your questions, or are you just going off at a tangent?'

He sighed pointedly. 'Could you just bear with me here for one minute, Brontë. This is important. How old are you?'

'Thirty-five, but I don't see—'

'You've got middle-child syndrome.'

'I've got what?' she said, starting to laugh, but he looked so perfectly serious she stopped. 'Okay, enlighten me. What's middle-child syndrome?'

He took a bite of his stroganoff, then sat back in his seat.

'The first child in a family is always the most anticipated and exciting for the parent so it's put on a pedestal, applauded for everything it does, and made a big fuss of. That's your brother, Byron. The baby of the family basks in the sentimentality of being the last child, so it's basically spoiled rotten. That's your sister, Rossetti.'

'And me?' she said, curious in spite of herself as she ate some of her poached salmon and found it to be every bit as good as he'd promised.

'Well, you… You were just there. Never getting the same praise as your brother and sister because having been through the toddler stages with Byron, your parents just expected you to learn how to do the same things he had, and Rossetti would be completely indulged because they knew she was going to be their last child. You were stuck in the middle, overlooked a lot of the time, so you grew up to suffer from severe low self-esteem.'

'I suspect your psychology is a bit skewed,' she observed uncomfortably, and he shook his head as he took a sip of water.

'Nope, it's pretty well documented.'

'But I don't feel inferior to my brother and sister.'

'*I'm the dummkopf of the family.* That's what you said,' he declared, 'and it bothered me—it bothered me a lot—that you should think being a nurse made you inferior to them.'

She wanted to tell Eli he was wrong, that she didn't feel at all inferior to Byron or Rossetti, but she suddenly realised she couldn't. It was hard not to feel inferior when her brother would telephone to say he was heading off to Hong Kong, or New York, or Tokyo, and it had been even harder to feel genuinely happy for Rossetti when their parents had bought her a plushy, three-bedroom flat while she was still living in rented accommodation.

'Okay, so maybe I do feel a little inferior to them,' she conceded, realising Eli was waiting for an answer. 'In fact, if I'm going to be completely honest, there were times, when I was growing up, and Byron won yet another collection of school prizes, and Rossetti appeared never to do anything wrong in my parents' eyes, I did wish I belonged to another family.'

'Knew it!' Eli exclaimed triumphantly.

'*But...*' Brontë continued determinedly. 'Every child feels inferior to its siblings at one time or another. I bet you did, too. I bet there were times when you thought, "Oh, to have been born into a different household, to some other family."'

'In my case, I would have been happy to have belonged to any family,' he murmured.

His face had that shuttered look again, she noticed. The look it always assumed when she strayed too far into something he clearly considered very personal, and she ate some more of her salmon, then slowly put down her knife and fork.

'Sounds like you had a rough childhood.'

'Water under the bridge now,' he said dismissively, but she sensed it wasn't, not for him.

'Eli—'

'Anyway, we're talking about you,' he interrupted, 'not me.'

*So back off, Brontë.* That's what he was saying. *Back off, keep out and don't probe.*

'Okay, if I accept you're right about me suffering from middle-child syndrome—which I don't,' she declared, seeing him roll his eyes as she picked up her knife and fork again, 'I fail to see how being a middle child explains my lousy track record with men.'

'It's obvious,' he said, attacking his stroganoff impatiently. 'If you're a middle child, with low self-esteem, that, in turn, makes you an easy target for fundamentally weak men who will walk all over you. You should be looking for a man who is self-confident, a man who is easy in his own skin.'

*Like you,* her heart whispered, and unconsciously she shook her head.

'Which is fine, in principle,' she argued back, 'but, surprising though it may be to you, even in the twenty-first century women don't tend to do the choosing, and if we see an example of the alpha male you seem to be describing there's precious few women who would go up to such a man, and ask *him* out.'

'But—'

'Eli, alpha men gravitate towards alpha women,'

she continued, 'so they are not going to be asking someone like me out.'

A flash of anger appeared in his blue eyes.

'What do you mean, "someone like you"?' he said irritably. 'What's wrong with you?'

'Look at me, Eli.'

'I *am* looking,' he protested.

'No, not like that,' she replied with exasperation. 'Look at me the way a man looks at a woman he's scoping out.'

'Yeah, right,' he said wryly, 'and like I want all my teeth knocked out.'

'I'll give you a pass on this one, I promise,' she insisted, 'just look at me.'

Obediently, he put down his knife and fork, steepled his fingers together and leant forward.

'Okay, I'm looking.'

And, dear heavens, he was, she thought, feeling a flood of heat surge across her cheeks as his gaze swept over her figure, then up to her face, lingering for an instant on her lips, and then returned to her eyes. That look would have melted a polar ice cap, and as for her… Thank heavens she was sitting so close to the fire, because if she hadn't been she didn't know how on earth she would have explained why she not only felt ablaze, but also looked it.

'Okay, tell me what you see?' she managed to ask.

No way on this earth was he going to tell her what he saw, he thought as he moved uncomfortably in his seat, all too aware of a painful tightness in his groin. No way was he going to say he saw a woman who wasn't even pretty, far less beautiful. A woman whose gold-flecked brown hair was currently sticking out at a wild angle, and whose nose was chapped and red at the tip because of the cold weather they'd been having. A woman who, for some insane, inexplicable reason he still wanted to lean forward and kiss.

She'd think he was crazy. *He* thought he was crazy, and he had done so ever since last night when, if she hadn't moved away—or maybe he did first, he honestly couldn't remember now—he would have kissed her for sure, and that would have been bad, seriously bad.

'I see...' He cleared his throat. 'I see a young woman with a very winning personality.'

'Oh, wonderful.' Brontë groaned. 'That's exactly what every woman wants to hear. *Not*.

Why don't you just offer me a paper bag to wear over my head the next time I go out?'

'Look, I phrased that wrongly,' he said quickly. 'What I meant—'

'It's okay, it's all right,' she said dismissively. 'I asked for your opinion, and you've given it. I know I'm not an alpha woman.' She thought about it, and frowned. 'Actually, I'm probably not even a gamma woman if there is such a thing, because I bet your first thought when you saw me on Monday night was, Ordinary, boring.'

It had been, he remembered. It had been exactly what he'd thought, and he'd been so wrong, so very wrong, just as he had been wrong to prolong their meeting after she'd bought her boots. Having almost made a complete fool of himself in Harpers & Stolins, by staring at her legs—and she had very good legs, he remembered, forcing himself not to look down and check them out again—he should simply have waved her farewell, and gone home, instead of urging her to spend more time in his company.

*So why did you?*

Because I wanted to prove to myself that last night—this morning—had been a temporary aberration, he thought ruefully. I wanted to see her outside of the station environment, so I'd realise she's just another woman, not in any way different or special, but the trouble is she *is* different and I don't know why.

'You're taking too long to answer, Eli,' she pointed out, and he felt a tide of heat creep up the back of his neck.

'Of course I didn't think you were boring,' he lied, and she shook her head.

'Yes, you did, and that's what I am. Brontë O'Brian. Memorable only because of my name, so thanks for your advice about dating only super-confident alpha men but, trust me, it ain't going to happen.'

'What about dating men who are all screwed up?' he said before he could stop himself, and Brontë laughed.

'Men like you, you mean?' she replied.

The minute the words were out of her mouth, she wished them back. What she'd said... It sounded almost—hell, it sounded *exactly*—as though she was hitting on him, and she hadn't intended to do that, and she opened her mouth to say something flippant, light, but the words died in her throat.

He had reached out and taken one of her hands

in his, and was staring down at it, his face a little wry.

'I certainly feel completely screwed up at the moment,' he murmured.

'Do you?' she said faintly, and he nodded.

'You see, the trouble is...' He met her gaze. 'I just can't figure you out at all.'

'I thought...' Oh, my heavens, but he was stroking the palm of her hand with his thumb, and she was having difficulty thinking, far less speaking. 'I thought we'd established I was easy to read. That I'm suffering from middle-child syndrome, and I'm ordinary and boring.'

'You're definitely suffering from middle-child syndrome, but you're anything but ordinary and boring. You're...' A rueful smile appeared in his blue eyes. 'You're one of a kind, Brontë O'Brian.'

*One of a kind.* Did that mean what she thought—hoped—it meant? Did it mean he found her desirable, not ordinary and boring at all, but actually desirable? She knew she should say something, but she didn't know what to say, didn't want to mess it up, and she cleared her throat, only to pause.

A log had shifted in the grate beside her, sending

a shower of snapping, crackling sparks spiralling up into the chimney, but it wasn't that which had caught her attention. It was the muffled, feminine laughter behind her and, as she glanced over her shoulder and saw the knowing smile on the face of one of the waitresses, she realised something she should have seen before. Something she must have been blind not to have been aware of, and a twist of pain tore through her.

Eli was flirting with her. Not because it meant anything, not because he was interested in her, but simply because he could, and she, like a poor sap, was falling for it, falling for his charm, just as all the other women in his life had. He was playing with her, and she had never played games when it came to relationships. For her it had always been all or nothing, and with Eli she knew as surely as she knew anything that if she didn't stop this right now it would simply lead to a broken heart, and she couldn't go that way, just couldn't.

Blindly, she pulled her hand out of his, and reached for her coat.

'I…I have to go,' she said through a throat so tight it hurt.

'But why?' he protested, confusion plain on his face. 'You haven't finished your meal—'

'I need... I have to go home to get changed,' she said, frantically searching under the table for her carrier bag from Harper & Stolins, desperate to get away from him.

'But, we're not on duty until ten-thirty,' Eli declared. 'We've another hour—'

'I need to shower as well,' Brontë said, looking everywhere but at him. 'I...I need to shower, and change, and get ready for tonight.'

She was already heading for the restaurant door and, without even counting them, Eli pressed some notes into the waitress's hand, and hurried after her.

'Brontë, what's wrong?' he demanded when he caught up with her in the street.

'Nothing...absolutely nothing,' she said with forced brightness. 'Thank you for the help with the boots, and the lovely meal. I'll see you tonight.'

'But, Brontë—'

*Don't come after me,* she prayed, as she walked quickly away from him down the High Street, the sound of her feet muffled by the now lying snow. *Please, please, don't come after me,* her heart cried as she dashed a hand across her cheeks, despising herself for her weakness and stupidity.

To her relief he didn't. To her relief, no familiar

voice called her name, no firm hand grasped her shoulder, but it didn't help. She might be able to lose herself amongst the other late-night shoppers who were thronging the street, but she couldn't shut out the mocking little voice in her head.

The little voice that whispered, *Fool, Brontë. You are such a fool.*

Eli rotated his shoulders wearily as Brontë drove carefully down Johnstone Terrace towards Stables Road. An RTC, Dispatch had said, and it wasn't the first such accident they'd been called out to tonight. In fact, from the minute they'd clocked on at ten-thirty, they'd scarcely had a moment's rest, not even for their customary coffee break. The now deeply lying snow, coupled with inexperienced city drivers unused to driving in such conditions, had meant accident after accident, and ED7's resources had been stretched to breaking point.

'The police are here,' Brontë observed as a flashing light suddenly appeared in the dark in front of them. 'Must be a serious one.'

It looked it, even from a distance and, when they got nearer, it looked even worse. The car had clearly skidded right across the road, but then it had hit a wall, and brought part of the wall down on top of it.

'I'm afraid the driver looks in a pretty bad way,' one of the policemen declared when Brontë and Eli walked towards him. 'According to her ID, her name is Katie Lee, aged twenty.'

And she hadn't been wearing a seat belt, Eli thought as he stared at the car. He could see the distinctive ringed crack on the windscreen which meant the young woman had 'bullseyed' it when her car had hit the wall, and there were tiny spots of blood and bits of hair embedded in the glass. Why didn't people learn—why did they never seem to learn?

'Dear heavens, the whole front of the bonnet is completely crushed,' Brontë whispered.

He could hear the shock in her voice, the appalled horror, but then she would only ever have seen people who had been brought in to A and E by ambulance, and not the situations they had come out of, as he always did.

'Watch out for broken glass, and bits of metal,' he replied, 'and if you see any oil, let me know. We don't want to be inside something that might go up like a fireball.'

'Right,' she said faintly. She glanced up at him. 'I'm okay, honestly I am. It's just…'

'The first bad RTC is always a bit of a shock.' He nodded. 'I threw up after my first one.'

She managed a smile but, when he smiled back, she looked away quickly and he bit his lip.

She'd scarcely said a word to him all night, and it wasn't simply because they'd been run off their feet. She was clearly edgy and uncomfortable in his company, and it was all his fault.

Why in the world had he virtually asked her whether she would consider dating him, he wondered, as he leant into the driver's seat to take the young woman's pulse, and heard the mobile phone lying on the car floor begin to ring. He still didn't know why he'd said it. The words had just come out, completely without warning, so thank heavens Brontë had taken the initiative, and ended the conversation. If she'd said, 'Okay, Eli, where do you want to go on this date?' he would have been in severe trouble because he'd made his no-dating pledge, and it still had a month to run.

*Oh, get outta here, Eli,* his mind laughed. *That isn't what's freaking you out. What's freaking you out is that for the first time in your life you feel all at sea with a woman, and it's scaring you half to death.*

'Katie…Katie, can you hear me?' Brontë said as she leant in through the broken window of the passenger door.

'My legs…my legs hurt so much, and my chest…' The young woman groaned as she tried to take a breath. 'It's like there's a big, heavy weight on it.'

'Pulse weak, BP falling,' Eli muttered as he swiftly wrapped a cervical collar round the young woman's neck to support it before they could attempt to take her out of the car. 'Looks like compound fractures tib and fib and possible pelvis fracture to me.'

It would have been Brontë's initial assessment, too, plus she strongly suspected severe internal injuries.

'Pethidine?' she suggested, and Eli nodded.

'Haemaccel drip, too,' he added. 'Is the heart monitor good to go?'

'Just about,' Brontë murmured. 'Okay, it's on.'

They both stared at the screen, then exchanged glances. The young woman's heart rate was erratic, extremely erratic, and the last thing they wanted, or needed, was her to go into VF within the small confines of the car, or, even worse, to go asystole. Asystole hearts couldn't be shocked. All they could do, if that occurred, was perform CPR, and affix an Ambu bag to try to keep the oxygen flowing to the young woman's brain, but casualties who presented at A and E with asystole rarely survived.

'Is there no way we can switch that off?' Brontë exclaimed as the young woman's mobile phone began to ring again, and she saw a message flash up on the display screen.

A message which read, 'Katie, can you call me? It's Mum.'

'Try to ignore it,' Eli replied.

Brontë gritted her teeth, and tried her best, but it was hard. Hard to see that message constantly flashing, and to know that unless they got Katie to A and E quickly she might never be able to reply. Hard, too, she thought as she glanced over at Eli's lowered head as he swiftly inserted a drip, to persuade herself that the man working so closely beside her meant nothing to her, that he was simply a colleague, but she was going to do it. She was going to distance herself from him no matter what it took. She had to for her own self-preservation.

'How are we going to get her out of the car?' she said when she'd finished attaching an Ambu bag. 'So much of it is crushed. Should we call for the fire brigade?'

Eli frowned, then, as Katie Lee groaned, he clearly came to a decision.

'We can't afford to wait. Her BP's going through the floor, and I think she's bleeding internally.'

'But...'

'We get her out, Brontë.'

And they did, by the simple expedient of Eli taking most of the weight of the roof of the car upon himself as well as Katie's legs and torso.

'Your shoulders are going to be black and blue tomorrow,' Brontë observed, seeing Eli wince when he straightened up after they had safely carried Katie into the ambulance.

'Nothing I can't live with,' he replied dismissively. 'And getting her to A and E is much more important than me getting a couple of bruises.'

He meant that, she knew he did. Even in the short time she'd been working with him she'd seen he would always go that extra mile for the people who needed his professional skills, and yet he wouldn't go one step for the women in his life. Was it simply because he truly didn't believe in love, or was there something else? She wished she knew. She wished even more that she didn't care about the answer because not caring would have been much, much safer.

'I'll be glad when this shift is over,' she said after she and Eli had taken the young woman to the Pentland. 'I'm shattered.'

'Me, too,' he replied. 'And yet, in a weird and horrible way, having to deal with all these accidents, fighting to keep people alive until we can get them to A and E… It's the sort of night we paramedics long for. Which—when you think about it—means we're either adrenaline junkies or must have hearts of stone.'

'I think it simply means if you've been trained to do difficult, complex procedures, the last thing you want is to be constantly presented with trivial ones,' Brontë said thoughtfully. 'I bet a plumber feels exactly the same. I bet he thinks, Wish I was out there fitting a challenging, state-of-the-art power shower system instead of unblocking yet another bog-standard sink.'

Eli stared at her in open-mouthed astonishment for a second, his eyes red-rimmed and shadowed with exhaustion, and then, to her surprise, he burst out laughing.

'Only you could think of that!' he exclaimed.

'But it's *true*,' Brontë insisted, and he laughed again.

'I know it is, but only you would say it out loud which is why I think I like…'

'Like what?' she asked, and saw him shake his head uncomfortably.

'Nothing. Not important.'

'But—'

'Isn't that Dr Carter?' he interrupted. 'Maybe she'll have some news for us about John.'

Helen Carter did, but it wasn't good.

'I'm really sorry, Eli,' she said unhappily. 'But it was Mr Duncan who did the rounds on Men's Medical this morning, and he created a bit of a stink about young John.'

'In other words, he kicked him out,' Eli declared with barely suppressed anger. 'He took one look at the boy's chart, refused to listen to any of the background information, and discharged him.'

'I think the words, "This is not a bed-and-breakfast establishment for down-and-outs" was used.' Dr Carter sighed. 'I did put a note on John's file, saying "exceptional circumstances"—'

'I'm sure you did, Helen,' Eli interrupted grimly, 'but Duncan's one of the old-school consultants, never does anything unless it's by the book. He just doesn't seem able to see that if we help people when they first present they'll need less medical attention in the future.'

'You did your best, Eli,' Dr Carter replied, her round face concerned.

'Yeah, well, it wasn't good enough this time, was it?'

And before Helen Carter could say anything else, he had walked away, and, with a quick glance of apology at Dr Carter, Brontë hurried after him.

'Eli, it wasn't your fault,' she said when she caught up with him in the street.

'And is that supposed to make me feel better?' he retorted. 'Peg *trusted* me, Brontë, and I failed her and the boy. Especially the boy.'

'At least you tried—'

'And a fat lot of use that did!' he snapped.

'Eli—'

'Just drive,' he said. 'I don't care where you go, but get me away from here because, if you don't, I swear I'll go back in there and find Duncan, and if I do...'

He didn't need to finish his sentence. Brontë waited only until he had snapped on his seat belt and she was off, driving down the Canongate, onwards past Abbey Mount, but, when she turned into Montrose Terrace, she heard him unclip his seat belt.

'I can't sit here,' he muttered. 'Stop the ambulance, Brontë. I need… I have to walk.'

Obediently, she drew up at the kerbside but, when he got out, and began walking up and down the pavement, his hands tight, balled fists at his side, she knew she couldn't simply sit there in the ambulance, watching him. No matter what he was, or what he had done, her heart bled for him. He had tried so hard to help John, had really cared about what happened to the boy, and she couldn't let him walk up and down the snow-covered pavement alone.

'Are you okay?' she asked tentatively as she approached him.

'What kind of damn fool question is that?' Eli exclaimed, then closed his eyes and, when he opened them, she could see regret there. 'I'm sorry, so sorry.' He gave an uneven laugh. 'Do you realise I've apologised to you more in the past four days than I have to anyone else in my whole life?'

'Maybe that's because you've never met anyone

quite as irritating as me before,' she said, hoping to at least provoke a smile, but she didn't. 'Do you want to go and look for John? We could try Greyfriars. He might have gone back to Peg.'

'Didn't you hear what she said when we took him away on Tuesday?' he said morosely. 'She said, "I don't want to see you back here again, young feller."'

'Yes, but she meant it kindly,' Brontë replied, hating to see him looking so defeated. 'She meant she hoped he was going on to a better life.'

'He won't see it as such. He'll see it as an order not to go back to Greyfriars. He could be anywhere, Brontë, and trying to find him…' He shook his head. 'There are so many homeless people in Edinburgh.'

'Perhaps he'll go home,' Brontë said without much hope, but feeling she had to try to give Eli some. 'Or if he doesn't go home, maybe someone from one of the charities might pick him up, and he'll be okay.'

'Brontë, just what planet are you living on?' Eli exclaimed angrily. 'He's more likely to become a crackhead, or a rent boy, than be *okay*.'

'I know—'

'You *don't*,' he replied, almost shouting at

her now. 'You have no idea what it's like to be out on these streets, alone, and friendless. No comprehension of the temptations, the quick-fix drugs that are offered to you which you're only too happy to accept because they allow you an escape for just a little while. Well, I *do*.'

'Eli—'

'Yeah, you were right about me,' he continued, his mouth twisting in a bitter parody of a smile as she stared up at him. 'I was out on these streets for a year, so I know *exactly* what's it like.'

His face was harsh and white in the lamplight, and she didn't know what to do, what to say, so she did the only thing she could think of. She put her hand on his arm in what she hoped was a comforting gesture and, for a second, she thought it might have helped. For a second, she thought he was actually going to cover her fingers with his own, then he pulled his arm free with a muttered oath.

'Eli, don't give up hope,' she insisted, thinking he was going to walk away from her. 'Every time we go out we can keep an eye out for him, and maybe we'll get lucky, maybe we'll find him.'

He could see sympathy and pity in her eyes and something inside him broke. He didn't want her sympathy, or her pity. He didn't want to feel the urge to put his arms around her, and the desire to have her hold him back. He was Elijah Munroe, who had never in his life needed anyone, and he wasn't about to start now. Somehow she was managing to get past his defences, and he had to stop it, end it, and he lashed out at her with the only weapon he had left.

'What—no questions about my time on the streets?' he jeered. 'Aren't you just longing for me to tell you all the gory, grisly details?'

She flinched, almost as though he had hit her, and he saw her eyes cloud with hurt.

'I thought…' He saw her shake her head. 'I hoped you might know me better than that. I would never intrude on something that's very personal to you.'

'Oh, of course, I forgot.' He nodded. 'Saint Brontë. Always the sympathetic, always the understanding. Little Miss Perfect.'

What little colour she had drained from her face.

'I'm not perfect—not by a long shot,' she said, and he could hear the tremble in her voice. 'I screw up just like everyone else, make mistakes as we all do—'

'You wanted to know whether I had any brothers or sisters?' he interrupted, hating to see the pain and bewilderment in her face, but it was preferable to sympathy and pity. 'I could have twelve for all I know because my mother dumped me in an orphanage when I was four. "I'll be gone for five minutes," she said, and then she disappeared, and I never saw her again.'

He waited for her to dish out the sympathy he knew he couldn't bear, because then he would be able to lash out at her again, but she didn't give him sympathy.

Instead, she said, 'Do you know why she left you?'

'Got fed up with me, I suppose,' he replied dismissively. 'Got fed up with having a whining brat hanging about her skirt all day, spoiling her life, and by all accounts I wasn't a very loveable child.'

Brontë didn't know whether he had been or not. All she knew was there was a world of hurt inside this man, a world of pain and self-hatred, and, though what he had said to her had been cruel, she had to swallow hard to subdue the tears she could feel clogging her throat.

'And your father?' she said hesitantly.

'I don't remember any man in the house, so I'm guessing I was a bastard.' He smiled. A tight smile that made her wince. 'Which is what you always believed I was, so you were right yet again, Miss O'Brian. Someone give the lady a prize.'

'How did you end up on the street?' she asked, longing to put her arms around him, to comfort him, but his rigid face told her that would be a very big mistake.

'I got fostered out a few times from the orphanage, but each family sent me back.' He shrugged. 'Too much trouble. Always getting into fights, always running away.'

'Because you were unhappy,' she protested. 'And what person—be they adult or child—wouldn't lose their temper, and keep running away, if they were suddenly uprooted from what they'd known all their lives and continually placed somewhere alien, strange?'

'Yeah, well, by the time I was fourteen I'd had enough of being passed around as the parcel nobody wanted, so I ran away. I lived for a year on the streets in Edinburgh until one of the priests from a night shelter found me sleeping beside their dustbins. At first I thought he was pervert, after my youthful body,' Eli continued with an ironic laugh, 'and the poor bloke ended up with a black eye, and a broken nose.'

Brontë didn't laugh—she couldn't—and she knew he wasn't laughing either, not inside. She could picture him oh-so clearly in her mind; how frightened he must have been, how desperate.

'He took you in?' she said, trying to keep her voice as even as possible, but knowing she wasn't succeeding.

Eli nodded. 'He taught me how to read and write, gave me books, paper and pens, and then he started taking me out on his pastoral visits, and I met all these people. People who were ill. People who were dying.'

'And that's when you decided you wanted to become a nurse.'

'Yes.'

She stared down the dark and empty street for a few seconds, then back at him.

'Eli, there are ways of tracking down your mother. Societies who specialise in finding missing people. You could—'

'And why the hell should I want to find her?' he interrupted, his face dark and angry. 'She didn't want me. If she'd wanted me she would never have dumped me.'

'Maybe she had a good reason,' she said uncertainly. 'Maybe she didn't want to give you up, but she didn't have enough money to take care of you, or maybe she was very ill, and knew she couldn't cope.'

His lip curled.

'Saint Brontë, always the understanding, always the forgiving. Doesn't it ever get tiring living up on that damned pedestal of yours? No wonder men walk all over you, use you like a doormat, because that's exactly what you are!'

She stared up at him, her face ashen in the moonlight, and, as a clock tolled the hour in the distance, he saw her swallow, then hesitantly pull the ambulance ignition keys from her pocket and hold them out to him.

'It's half past six,' she said. 'Our shift's over so, I think... As I live just round the corner... Would you mind very much driving the ambulance back to the station for me?'

'Not a problem,' he said dismissively, and she backed up a step.

'It's just...I think...I would like to go home now,' she said.

'You do that,' he replied.

But, as he watched her trudge away through the

snow, her shoulders hunched against the biting wind, he had to clench his hands together to stop himself from running after her. To stop himself from getting down on his knees and begging her to forgive him for the awful, unforgivable things he'd said.

It's better this way, he told himself when she turned the corner and he couldn't see her any more. Much better, he insisted, closing his eyes tightly against the snow which was beginning to fall again, and if he didn't feel that way at the moment he eventually would. Time, as he had discovered, could make you accept almost anything.

# CHAPTER FIVE

*Friday, 10:14 p.m.*

'Okay, I want to know what happened, and I want it in words of one syllable,' George Leslie declared, his normally amenable face tight and angry.

'I'd be more than happy to give you an answer if I knew what you were talking about.' Eli smiled as he rubbed down the windscreen of his ambulance, then tossed the chamois leather into the glove compartment.

'Ms O'Brian.'

Eli's smile disappeared in an instant. 'What about her?'

'Don't play the innocent with me,' his boss retorted. 'Three more shifts—that's all you had to keep a lid on your blasted temper for. Three shifts to be pleasant, and civil, but, no, you had to go and screw it up.'

'She's been complaining about me, has she?'

Eli observed, his face an unreadable mask, and George Leslie grimaced reluctantly.

'She phoned me this afternoon to ask if she could be assigned to another ambulance because she felt she had upset you. Complete nonsense, of course,' George continued with a pointed glare at Eli. 'I know perfectly well if there was any upsetting going on you were at the root of it.'

'She thinks she's upset me?' Eli said incredulously, and his boss nodded.

'Frankly, I didn't believe a word of it either, but—'

'Who are you teaming her with?' Eli interrupted, and George Leslie threw his hands up with irritation.

'As I told Ms O'Brian, I can't team her with anyone else unless I do some massive alterations of people's shifts which will go down like a lead balloon with the other paramedics, and for three shifts it's just not worth the hassle, so she's stuck with you.'

'I see.'

'I don't think you do,' George Leslie observed. 'Eli, I want whatever has caused this friction sorted. She seems a pleasant enough person, not at all like a normal number cruncher—'

'She isn't.'

'—so can you *please* try not to rub her up the wrong way?'

'George—' Eli began, only to clamp his mouth shut.

The bay door had opened, and Brontë had stepped hesitantly onto the forecourt. George had seen her, too, and his eyebrows snapped down.

'I don't want to have this conversation with you again, Eli,' he declared in a hissed undertone. 'Understand?'

Eli didn't as his boss walked away. He didn't understand why Brontë hadn't simply told George he was impossible to work with. He fervently wished she had as he watched her square her shoulders as though preparing for his next onslaught. He had said such terrible things to her at the end of their last shift, things which still made him feel guilty, and he didn't want to feel guilty. He wanted his easy, uncomplicated, carefree life back again, not the woman who was so unsettling it.

'Cold night again,' Brontë declared hesitantly as she drew level with him. 'Minus twelve, according to the forecast, and there must be about six inches of snow on the roads now.'

'Yes,' he replied.

Heavens, but she looked so tired. He could see dark shadows under her eyes as though she hadn't slept well. He had slept badly, too, tossing and turning, unable to forget the look on her face before she'd turned and walked away from him.

'Hopefully we won't be as busy as we were last night,' she observed.

'Hopefully not.'

'Eli—'

'Brontë—'

They'd spoken in unison and, though he knew it was wrong, he took the coward's way out.

'You first,' he said, and saw her bite her lip, then take a deep breath.

'I saw you talking to George,' she said. 'I did try to get him to change the roster—to assign me to someone else—honestly I did, but he said it would be too difficult. I'm sorry.'

She looked it, too, he thought with mounting irritation. She looked as though she wanted to be anywhere but here, talking to anyone but him, and his guilt morphed into a much more convenient anger.

'Why the hell didn't you just tell George the truth?' he demanded. 'That I was rude and obnoxious to you last night, and you can't stand working with me?'

She looked confused.

'But that wouldn't have been the truth,' she said. 'I overstepped the mark—intruded into your private life—and you had every right to chew me out.'

He stared at her in disbelief. Surely she couldn't truly believe that? She must be being sarcastic, but there was nothing in her grey eyes except genuine regret, and he swore in exasperation.

'Dammit, Brontë. Must you always be so understanding, so accommodating, so…so damn *nice*? After what I said to you this morning… You should be calling me a low-life scumbag, and slapping my face!'

To his surprise, a hint of a smile appeared on her lips.

'I didn't know you were into S and M, as well as serial dating.'

'S and M…?' His mouth opened and closed soundlessly for a second, then he gazed heavenwards with frustration. 'Brontë, you are *impossible*.'

'Probably.' She nodded. 'You know, I'm really clocking up the labels since I started working with you. I always knew I was boring and ordinary—'

'You are not!'

'And now you've added middle-child syndrome, low self-esteem, doormat and impossible to my labels.'

'The doormat crack,' he said uncomfortably. 'That was totally out of order.'

'It would have been if I don't have a horrible suspicion you're right,' she replied ruefully. 'So, in a way, you did me a favour, and this morning I made a pledge. No more nice-girl Brontë. It's kick-ass Brontë from now on.'

'I'm glad to hear it,' he said, meaning it, 'but that doesn't absolve me from guilt. I was the jerk last night. Me, not you. *Me*.'

'But—'

'I was angry,' he interrupted. 'I know that's no excuse, but I was angry with Duncan for being so stupid. Angry with society for turning its back on youngsters like John, and...' He met her gaze. 'I was angry with you for somehow getting me to talk about myself, my past. I don't do that, you see. Not ever.'

'Maybe you should,' she said gently, her large grey eyes soft.

'The past is past,' he replied dismissively. 'It does no good to resurrect it, dwell on it, because nothing can be changed, altered.'

'No,' she agreed. 'But sometimes the past needs to be faced, to be come to terms with, so it can't hurt you any more, and you can move on.'

'Like you getting over your lousy track record with men, and conquering your low self-esteem?' he suggested with a glimmer of a smile, and she smiled back.

'Exactly,' she said, then she tilted her head speculatively to one side, and he saw a glint of laughter appear in her eyes.

'What?' he asked warily.

'I'm just wondering whether you still wanted me to slap you?'

'Depends on how big a punch you pack,' he replied.

She laughed, and he tried to laugh, too, but all he could wonder was, How had she done that? One minute he had been so angry with her, and then the next… And he wanted to stay angry with her, because being angry was safe. Being angry meant she couldn't affect him, and being angry with her was infinitely preferable to the undeniable relief he had felt when he'd seen her smile.

*Responsible, protective,* his mind whispered, and he tried to crush down the nagging little voice, tried to make it go away, but it wouldn't go away, and when he heard the bleep of their MDT he beat Brontë into the ambulance by a long country mile.

'What have we got?' she asked as she joined him.

'LOL, 22 Jeffrey Street, Violet Young,' he replied. 'Neighbour can see her lying in the hallway, apparently unconscious, but can't get in because the front door is locked.'

'Well, I don't think that's even remotely funny,' Brontë muttered as she drove out of the bay.

'Sorry?' Eli replied in confusion, then the penny dropped and he laughed, properly this time. 'LOL doesn't mean "laughing out loud," as it does in chatspeak, Brontë. It means "little old lady."'

'Then why didn't the display say that?' she protested, and he smiled.

'Yeah, and like A and E doesn't have a language all of its own?'

He was right, it did, Brontë thought, trying not to smile back too fulsomely, but it didn't work, and she groaned inwardly.

What had happened to the pledge she'd made

so sincerely this morning? The pledge the new kick-ass Brontë had made to be pleasant and yet detached, friendly and yet slightly aloof. All that resolve had disappeared in an instant under the spell of a pair of deep blue eyes and a killer smile.

Distance, she told herself. What she desperately needed was to maintain some distance, but how did you distance yourself from a man when your wayward heart kept letting you down? When you discovered the more you found out about him, the more you wanted to hold him, not because he was handsome, not because he could be so charming, but because of the pain and hurt you now knew he kept so carefully hidden from the world. She was a lost cause, and she knew it, but of one thing she was certain. He was never going to find out how she felt. Not ever.

'Lot of houses,' she said, deliberately changing the subject as she turned into Jeffrey Street, and saw the row of buildings stretching far ahead of her. 'Let's hope the numbers run concurrently.'

They did but, even if they hadn't, the neighbour who had called 999 was out on the snow-covered pavement and waved them down.

'I'm so glad you're here,' she declared the

instant Eli and Brontë got out of the ambulance. 'Mrs Young... I thought she was just staying indoors today because of the snow, but her daughter phoned me half an hour ago. Said she couldn't get a reply when she dialled her mother's number, so I went round and...'

'She didn't answer when you rang the doorbell?' Eli said, and the neighbour shook her head.

'I looked through the letter box, and she's just lying there, in the hall, and...she's not moving.'

That the neighbour was deeply upset was clear, and Eli smiled at her encouragingly.

'Why don't you go back indoors now?' he suggested. 'We'll take it from here.'

The woman looked from Eli to Brontë, then back again. 'Are you sure? I don't want to leave my kids alone for too long, even though they're in bed, but if I can help...?'

'You've done more than enough already,' Eli insisted, and relief appeared instantly on the woman's face.

It was a relief Brontë would have felt herself if she'd been in the woman's shoes. There was just something about Eli that inspired confidence. Something which suggested if anyone could make things right he could. He was such an enigma, such a conundrum. Consummate professional, Good Samaritan and serial womaniser. Which of them was 'the real' Elijah Munroe? Maybe they all were. She wished she knew.

*No, you don't,* her head reminded her. *You're going to keep your distance, remember?* But, as she watched Eli crouch down to look through Mrs Young's letter box, she knew her pledge was a forlorn one.

'Can you see her?' she asked, and he nodded.

'She's at the very end of the hallway, and from the odd way she's lying, I'm thinking possible broken arm or leg, though it could be a CVA. The hall light's on which would suggest she collapsed—or fell—sometime this evening rather than earlier in the day.'

'Which is good news, isn't it?' Brontë replied, but Eli wasn't listening to her. He was running his hands over the door frame, and she stared at him incredulously. 'You're not thinking of trying to break down that door, are you?'

'Somewhat horrifyingly, you don't have to be superman to do it,' he murmured. 'That's just a normal front door key lock, and if you get the pressure dead centre, they break pretty easily.'

'They do?' Brontë gulped, thinking of the

exact same lock she had on the front door of her own flat.

'A deadlock is much better any day of the week,' he observed, 'and if you add a bolt to the deadlock, you're even safer. To be completely secure you could put a bar across your front door, though that would mean if you collapsed behind it we wouldn't be able to get in to resuscitate you, but I guess you'd be a safe corpse.'

'Right,' Brontë replied, making a mental note to phone a joiner the minute she finished this shift.

That mental note became a certainty when she saw Eli stand back, take a deep breath, then hit the door squarely with his foot and the lock gave way instantly.

'I'm impressed,' she declared, then shook her head. 'No, I'm not. I'm horrified you can do that so easily.'

'It helps if you take a size eleven.' He grinned, then swore when, out of nowhere, an Alsatian dog suddenly appeared at the end of the hallway, its teeth bared in a deep growl. 'Oh, wonderful. Why didn't the neighbour tell us she had a dog?'

'What do we do now?' Brontë whispered, only to wonder why she was whispering because it wasn't as though the dog would understand what she was saying.

'We'll need to phone the cops. Get them to send out one of their dog handlers.'

'But Mrs Young...' Brontë protested. 'The longer we wait...'

'I know, but do you really want to take on Fang there?' Eli demanded, and Brontë stared at the dog.

She'd grown up with dogs. There had always been at least two at home when she, and Byron and Rossetti, had been growing up, and, though none of them had been quite as big as the Alsatian standing in the hallway, the dog's eyes looked more frightened than angry.

'Good dog,' she said softly, taking a step into the house. '*Nice* dog.'

'Are you *crazy*?' Eli hissed, catching hold of her arm, and the dog's growl became a snarl.

'I know what I'm doing,' Brontë said out of the corner of her mouth. Or, at least she hoped she did as she saw the dog crouch, and the hackles on its back rise. 'Let go of me, Eli.'

'But, Brontë...'

'*Lovely* doggy, we're not going to hurt you, or your mistress,' she continued, keeping her eyes fixed on the dog as she advanced another step. 'I know you're frightened, and we're strangers in your home, but we're here to help.'

The dog sat down. Its lip was still curled back over its teeth, and it hadn't taken its eyes off her for a second, but at least it no longer looked to be in 'I'm going to spring and rip out your throat' mode, and the snarling had lessened to a low, warning growl.

Faintly she could hear Eli muttering, and guessed he was telephoning the police to ask them to send out a dog handler. Or for someone to sweep up her remains, she thought wryly as she walked forward another step, watching the dog the whole time.

'*Nice* dog, *friendly* dog,' she crooned. 'Everything's going to be fine, just fine.'

She risked a quick glance at Mrs Young. She thought the elderly lady's chest was rising and falling slowly, and for a second she thought she saw her eyes flicker open, but she couldn't be sure. She could also see a dark blue plastic toy shaped like a bone, and an idea came into her head. It wasn't much of an idea, but it was better than nothing, and nothing was what she had at the moment.

Slowly, she bent down, and picked up the toy.

'Want to play fetch?' she asked.

The dog's tail thumped once against the floor, then stopped. It was clearly torn between a favourite game, and guarding its mistress, but the door to the sitting room was open, and if she could just lure the dog in that direction…

'Brontë…'

'Shut up, Eli,' she said, hoping her tone sounded more like an endearment than an order to the watching dog. 'Want to play fetch?' she repeated, waving the toy bone backwards and forwards slowly and saw the dog's head follow her movements.

It was now or never, she realised, and, taking a deep breath, she shouted, *'Fetch,'* then threw the toy as hard as she could into the sitting room. The Alsatian was off like a flash, following it, but Brontë was faster. The minute the dog was in the sitting room, she pulled the door shut, then leant against it, breathing hard, keeping her fingers tight round the handle.

'Was that the police you were phoning?' she asked as Eli brushed past her towards Mrs Young.

'They'll be here in three minutes,' he answered.

His voice was tight, cold, but she ignored that.

'I hope they make it faster,' she said instead, as a cacophony of growls, and scraping paws, began to fill the air, and the sitting room door juddered beneath her fingers. 'My parents used to have a dog who could turn any door handle with its teeth.'

Eli didn't reply. He was already setting up their heart monitor, and, though she wanted to go and help, the thought of one very angry dog suddenly bursting out of the sitting room was not an appealing one.

'How is Mrs Young?' she asked.

'Heart rate not too bad given her age, pulse a little weak, and my guess is her right leg is broken,' he answered.

And Eli clearly needed help, not her standing with her fingers clamped tightly round the sitting room door handle, so it was with a huge sigh of relief that Brontë greeted the burly, uniformed figure who suddenly appeared at the front door.

'I hear you have a dog problem?' The man grinned.

'You could say that,' Brontë admitted as the sound of barking and snarling intensified from inside the sitting room. 'Could you hold on to this door for me, and not open it until my colleague and I have taken the dog's owner to the Pentland?'

'No problem,' the dog handler replied, and, as soon as his hand had replaced hers on the door handle, Brontë hurried to Eli's side.

'What do you want me to do?'

'Get an Ambu bag on her, and then we go,' he declared. 'She's very, very cold, which means she's been lying here for quite some time.'

'She has a name, young man,' Mrs Young murmured faintly, and Eli smiled down at her.

'So, you're awake, are you, Mrs Young?'

'No, I just talk when I'm unconscious,' she muttered. 'Of course I'm awake. I just don't seem to be able to *stay* awake.'

Brontë's eyes met Eli's. Hypothermia. There was a very strong possibility that Mrs Young was suffering from hypothermia as well as a broken leg.

'I'm afraid we're going to have to take you to the hospital, Mrs Young,' she said gently and saw a flicker of alarm cross the old lady's face.

'But what about Bubbles? You'll have to phone

my daughter, tell her to come and take care of him because I can't leave Bubbles on his own.'

The Alsatian who was currently trying to tear down the sitting room door was called *Bubbles*? Brontë glanced across at Eli to share the joke but, oddly enough, he didn't seem to find it nearly as amusing as she did.

'My colleague is just going to put this Ambu bag on you, Mrs Young,' he declared, 'and then it's off to the hospital for you. And, yes, I'll get someone to phone your daughter about Bubbles,' he added as the old lady began to protest, 'so stop worrying about him and start thinking about yourself.'

A small smile appeared on Mrs Young's lips, and she transferred her gaze to Brontë.

'Is he always this bossy, dear?' she said, and Brontë nodded.

'I'm afraid so,' she relied. 'In fact, he's notorious for it so, if I were you, I'd just give in gracefully.'

Mrs Young did and, once the Ambu bag was in place, they lifted her into a carry-chair, and quickly out to the ambulance.

'I'd say she's going to be okay, wouldn't you?' Brontë said to Eli after they'd delivered Mrs

Young into the waiting arms of the A and E staff. 'I know she's elderly, and if you break a leg at that stage in your life it can really knock you for six, but her heart rate was strong, and she seems a spunky, never-say-die lady.'

'I'd bet money on her being back home within the month,' Eli replied as they walked through the waiting room to the exit.

'Her daughter seemed nice, too,' Brontë continued. 'And I'm so pleased she's going to look after Bubbles. I know you didn't take to it,' she added, seeing Eli's jaw tighten, 'but it was only trying to protect its mistress.'

Eli said nothing.

'I know it looked a bit scary,' Brontë continued, 'but you have to look at it from its point of view. Its mistress wasn't moving, and we were strangers in its home.' Eli's silence was now positively deafening, and Brontë blew out a huff of impatience. 'Okay, you're clearly itching to chew my head off, so why don't you just do it, and get it over with?'

Eli waited until they were safely out on the street, and then he rounded on her.

'What you did—with that dog,' he said through

clenched teeth, 'was either the bravest thing I've ever seen, or the stupidest.'

'And my guess is you're favouring the stupidest,' she said with a smile, but he didn't smile back.

'I said we should wait for the dog handler,' he declared, 'but did you listen to me, obey me? No, of course you didn't—'

'Eli, I know dogs—'

'You know every dog in the world?' he flared. 'Whoa, but that must make you pretty unique.'

'I didn't mean I know every dog,' she protested, 'but I weighed up the situation, took a calculated risk—'

'One which almost had me flat out on the floor beside Mrs Young with a heart attack!'

He wasn't joking. She could tell from his taut face, and angry eyes, he wasn't joking, and she bit her lip.

'Look, I'm sorry if you were worried, but I was watching the dog's eyes, and I thought it seemed more frightened than anything else.'

'You thought—you *thought*!' He muttered something unprintable under his breath. 'Brontë, you had no way of knowing what that dog might do. It could have bitten you very badly, torn your hand off—'

'But it didn't,' she insisted. 'I'm not a fool—'

'Do you really want me to comment on that?'

'—and if it had looked completely out of control I would have waited for the dog handler. I *would*,' she insisted as he shook his head at her, 'but with dogs it's generally their eyes you go by. Their eyes, and their body language.'

'And that's the biggest load of rubbish I've ever heard,' he retorted. '*All* dogs are unpredictable. *All* dogs can turn into killing machines, and the bigger they are, the more damage they can inflict. I remember once, when I was living on the streets…' He closed his eyes, then opened them again. 'Let's just say I don't even want to think about what happened to the poor bloke, far less talk about it.'

She stared silently at him for a second, and when she spoke her voice was low, contrite. 'I'm sorry…I didn't realise…. Were you really that worried about me?'

It took him all his self-control not to reach out, grab her by the shoulders and shake her. Worried? His heart had practically stopped when she'd walked towards that dog, looking so small, so vulnerable, but he didn't say that.

'If you put your hand on my chest you'll feel my heart still going like a train,' he said instead.

A tinge of colour appeared on her cheeks, and she laughed a little shakily.

'I'll take your word for it,' she said, then cleared her throat. 'I really am sorry. I promise I won't ever do something like that again.'

'Is that a solemn "cross your heart and hope to die" promise?' he demanded, and when she hesitated, as though having to consider his question, his eyebrows snapped together. *'Brontë!'*

'Just kidding,' she said with a smile. 'It's a promise.'

He fervently hoped it was as they went back to their ambulance. He didn't ever want to go through another ten minutes like that ever again.

*Protective, responsible,* the irritating little voice in his head whispered again, and he tried to shut it up, to tell the annoying voice he would have felt the same concern for anyone, but the little voice simply laughed and, as it did, the answer suddenly came to him. An answer that was so blindingly obvious in its simplicity he wondered why he hadn't thought of it before.

Why in the world was he angsting like this? he wondered as Brontë drove away from the hospital. He was attracted to her, and he was pretty sure she was attracted to him, so all he needed to do was ask her out. His no-dating pledge only had three weeks left to run anyway, so what did it matter if he cut it short? He would simply ask her out, they'd go out on a few dates, become lovers and then, once he'd got her out of his system, he'd move on as he always did, with the problem solved, the angsting over.

'What's funny?'

'Funny?' he repeated, glancing across at Brontë in confusion.

'You're smiling,' she observed, 'so I just thought… If you've got a good joke I could sure do with hearing it.'

The joke was on him, he thought, because the solution had been there all the time, staring him right in the face, and yet for some unaccountable reason he hadn't been able to see it. He could see it now. He was back on familiar territory now, and it felt good.

'Brontë, I was wondering,' he began, only to groan when their MDT bleeped into life.

'"Code amber. RTC. Male aged thirty-two,"'

Brontë read. '"Junction of College Street and Nicolson Street. Two cars involved. One casualty with whiplash. Other driver uninjured. Police in attendance."' She frowned. 'Sounds like a possible shunt to me, if only one person is hurt.'

'My guess is it's a payment point,' Eli replied, wishing the casualty to the farthest side of the moon as Brontë turned onto the North Bridge.

'A what?' she asked.

'In RTCs almost half of those supposedly suffering from whiplash are actually people who are attempting to get the insurance money out of the other driver,' Eli explained. 'And the "payment point" is the part of the neck they keep pointing to, and declaring they're in agony.'

'We didn't call them "payment points" in A and E at the Waverley,' Brontë declared. 'We called them PITAs.'

'And I don't need my nursing diploma to guess what that acronym stands for.' Eli laughed. 'And it just about sums these jokers up. They waste our time, the police's time, and put the other driver through hell, and all for money.'

'Isn't that why RTAs are now called RTCs?' Brontë observed as she headed for Nicolson Street. 'Because some smart lawyers figured out if the police described a car crash as an accident, their clients would be off the hook?'

'Yup.' Eli nodded. 'So now we have road traffic collisions, but we still get the payment-point brigade.'

The policeman who was waiting beside the two cars when they arrived clearly thought Eli's diagnosis was the correct one.

'He's kicking up a real racket—swears he's in total agony,' he declared, 'but the idiot keeps forgetting which part of his neck is supposed to be hurt. It's the poor woman in the other car I feel sorry for. Her car got the worst of it, and she can't stop shaking.'

She couldn't, and it took Brontë a good fifteen minutes to calm the woman down, and persuade her there was nothing she could do here, and she really should phone for a taxi and go home.

'How's our whiplash patient?' she asked when she was eventually able to join Eli by the other car.

Eli rolled his eyes heavenwards. 'What do you think?'

'The Pentland?' she said, and he nodded.

'I know it goes against the grain to transport someone we believe is faking it to hospital,' he replied, 'but there's always the possibility—no matter how remote—that an X-ray will reveal an injury I've missed.'

Eli was right, but Brontë couldn't help but think, as she drove towards the hospital, that it was hard to feel sympathetic towards a 'casualty' who, despite constantly protesting he was in the most appalling pain, still managed to make a dozen phone calls on his mobile phone.

That journey—and their patient—pretty well set the tone for the next few hours. The sensible Edinburgh citizens might have decided after last night's road chaos to stay home and keep safe, but there were still enough of the young, and the just plain cavalier, out on the streets, to keep Eli and Brontë busy until well after four o'clock.

'And it's snowing again,' Brontë said irritably as she drove down the Canongate, and had to switch on her windscreen wipers to clean her screen. 'Which no doubt means even more idiots will decide it might be a "fun" idea to take to the roads, and see how far they get before they skid and hit something.'

'Or someone.'

She knew who he was thinking of, and she glanced across at him quickly.

'I've been looking for him, Eli,' she said. 'When I've been driving along… Every time I pass a group of kids, or some homeless people, I've been looking for John.'

'Me, too,' Eli replied. 'It's been so cold these past couple of nights, Brontë, so very, very cold. I keep hoping he's got a room in one of the shelters, but there's so little accommodation available.'

'Do you think Peg will be all right?' she asked tentatively, and saw him force a smile.

'Peg's a pretty tough cookie. If anyone can survive, she can, but there are so many homeless people out there. When I see elderly people like Mrs Young, I think of the long lives they've had, the happy memories they must be able to look back on, but young addicts, young alcoholics…' Eli shook his head. 'All I can think is, How did you get lost, how did you lose your way? Other kids your age will have a life, a family, friends, but you… Your life is going to be so short. So very, very short.'

'My brother, Byron, has this theory,' she observed. 'He says people make their own choices in life.'

'That's true to a certain extent,' Eli replied, 'but what power did we—as a society—give these young people to enable them to make any kind of choice?'

'Eli—'

'Sorry—sorry,' he interrupted with a rueful smile. 'It's one of my pet hobby horses, and I shouldn't inflict it on you.'

'You're not inflict—'

'Coffee. I need a coffee,' he insisted, 'so let's head for Tony's.'

He saw her hesitate, then take a deep breath.

'Actually, I've been thinking about what you said before—about me not having met or spoken to any of the other paramedics at the station,' she said in a rush. 'You made a very good point, so I think we should take our break at the station tonight.'

Damn, but that was the last thing he needed, to be surrounded by his colleagues, when what he wanted was some privacy so he could ask her out.

'Good idea, in principle,' he observed lightly, 'but Friday nights... They're always busy, everyone flat out, so there's little likelihood there will be anybody for you to talk to.'

'Oh.'

She didn't look happy, but he wasn't about to

waver, and so he smiled what he hoped was his best encouraging smile.

'Tony's?' he said, and thought he heard her sigh with resignation as she nodded, but he couldn't be sure.

Brontë was sure. Brontë didn't want to go anywhere near Tony's tonight. She wanted to be safe amongst a crowd. She wanted other people to distract her, not to be sitting alone in an ambulance with Eli, but she would rather have stuck a fork in her eye than admit that to him.

Be pleasant, and yet detached, she reminded herself when she pulled up outside Tony's. That's all you have to be. Pleasant, detached, and slightly aloof. How hard could that be?

Damned hard, she thought as he got out of the ambulance, then turned to look at her with the smile which always made all rational conversation disappear instantly from her brain.

'Cappuccino and doughnut?' he said.

She thought about it. 'A cappuccino, for sure, but tonight I want a hamburger. A big, juicy hamburger, with lots of onions.'

His eyebrows rose, but he didn't say anything until he returned to the ambulance with their orders.

'Never thought I'd see you eat one of those,' he observed, as she bit into the hamburger, then closed her eyes, clearly relishing the taste.

'Yeah, well, you're corrupting me.'

'Am I?'

It could have been a completely innocuous observation. It could have been the kind of joking thing a friend would have said, but it wasn't, and she knew it wasn't. His voice was suddenly low, velvety, within the silence of the ambulance, and the atmosphere had changed—she could feel it, sense it, but this time she knew what he was doing; this time she was prepared, and she opened her eyes, and met his gaze full on.

'Only for hamburgers,' she said.

'Really?'

His voice was teasing, liquid, and she felt her heart pick up speed.

'Really,' she replied firmly, wishing he would just let it go, would stop, but he didn't.

Instead, he looked at her over the rim of his coffee, his eyes dark and oh-so blue. 'Care to make a wager on that?'

She took another mouthful of hamburger, and swallowed it with difficulty, all too aware her heart rate had now gone into overdrive.

'Nope,' she replied. 'Not interested, not a betting woman.'

'Then maybe I can offer you something you *will* be interested in,' he said softly.

This wasn't merely flirting, she realised as her eyes met his, and what she saw there made her stomach lurch and her pulses race. This was something more, and, though part of her wanted to tell him to stop, to say nothing else, the other part—the weak, traitorous part—wanted to hear what he was going to say, and it was that part which won.

'I'm listening,' she said.

He put down his coffee.

'Brontë, I know you think I'm smug, and arrogant—'

'And a bit of a prat at times,' she interrupted. 'Don't forget the "bit of a prat."'

'It's engraved on my heart,' he said lightly, though she could see she had rattled him. 'But the thing is… You're a very special woman, Brontë.'

He was going to ask her out. He was going to ask her out, and, though she knew exactly what she should say in reply, knowing it didn't mean she was necessarily going to do it.

'Go on,' she said, taking a slow sip of her coffee more to buy herself some time than from any real thirst.

'Without being vain,' he continued, 'I think you like me, too, so I was wondering whether you'd like to come out to dinner with me before our shift tomorrow?'

She stared down at her coffee, then up at him.

'To dinner?' she repeated slowly. 'Now, are we just talking dinner here, or are we talking something more?'

He smiled his killer smile.

'Oh, come on, Brontë, we're both adults, and I think you and I could really have some fun together.'

Some fun together. Not a proper relationship, not the possibility of that relationship ever leading to something more permanent, but just some 'fun', and her heart constricted with pain and disappointment as she stared at his handsome face. Pain and disappointment which were very quickly overtaken by anger. A blazing furious anger with herself for yet again being so naive to hope he might have offered more, and an equally furious anger with him for believing she had so little self-worth she would even consider settling for what he usually offered the women in his life.

Carefully, she put her coffee down on the dashboard in case she was tempted to do something rash with it.

'And how long would this "fun" last?' she asked, keeping her voice neutral with difficulty.

He blinked. 'Sorry?'

'I'm just wondering, you see,' she observed, 'whether I'd get the usual two months with you, or whether I might get real lucky and be allowed a little bit longer.'

His dark eyebrows snapped together.

'I don't think that comment was necessary,' he replied icily, and she forced herself to shrug.

'I would have said I was just being realistic,' she replied, 'because, you see, what you call "fun," results in a hell of a lot of heartbreak for a lot of women, so it's a road I'm rather reluctant to travel down.'

'I have never in my whole life broken any woman's heart!' he exclaimed. 'I've always been upfront, never promised I'd stick around for ever.'

Why could he not see it? she wondered. Why did he keep on denying what was so obvious to her?

'Eli, are you blind, or stupid, or both?' she demanded. 'Can't you see—don't you understand—that it doesn't matter how upfront you think you're being? I doubt if there's a woman alive who, when you've asked her out, hasn't thought, This is for keeps, this is going to last. No matter what you say, no matter how hard you try to persuade them you're only dating them for "fun," they think, Me. He's going to settle down with me because I can change him.'

'I don't believe that,' he retorted, and she shook her head at him.

'That's because, for all your talk, you know *nothing* about women. You just take and take, and give nothing of yourself. Dating—sex—it's all a no-risk game to you, isn't it? Don't let anyone close, don't let anyone into your mind, and heart.'

'And you're such an expert on dating, are you?' he snapped, and hot, furious colour stained her cheeks.

'At least I *tried*!' she exclaimed. 'At least when I went into a relationship, I went into it wholeheartedly not thinking I'll just have some "fun," then dump the guy. Okay, so maybe my relationships didn't work out, and I ended up getting my face pushed in the mud, and my heart trampled in the dirt—'

'A bit difficult to have those two things done to you at once,' he interrupted sarcastically, and she moved beyond anger into incensed.

'Must you make a joke out of *everything*?'

'Look, listen—'

'No, *you* listen,' she broke in, her grey eyes blazing. 'Relationships, love—everything's one big game to you, and do you want to know why? It's because—deep down—you're a coward. You play at everything, never letting anyone get close to the real you, hiding behind this...this supercool image when, in reality, you're just a coward. I know your mother hurt you very badly—'

'Leave my mother out of this.'

His voice was low, dangerous, but she'd started and she couldn't stop.

'And maybe her leaving you resulted in you having trust issues—'

'You're saying I've set out to make every woman pay for what my mother did?' he exclaimed in disbelief, and she exhaled with exasperation.

'Of course I'm not saying that,' she replied. 'What I'm saying is I think you're scared to get too close to anyone in case you get hurt again.'

He shook his head impatiently.

'That is the biggest load of psychobabble I've ever heard.'

'It makes just as much sense as you telling me I have lousy taste in men because I have middle-child syndrome,' she countered, and he crushed the paper bag which contained his uneaten hamburger between his fingers.

'I've had enough of this conversation.'

'You were the one who started it,' she pointed out, and he rounded on her.

'Then I'm the one who's finishing it!'

'Fine.' She nodded. 'Throw a hissy fit, sit there in a snit, because you don't like the truth when you hear it, but you can do it on your own.'

'Where the hell are you going?' he demanded as she opened the driver's door, and got out.

'Back to the station where the air is less toxic.'

'Don't be ridiculous,' he declared. 'You can't walk back through The Meadows at half past four in the morning. Heaven knows what weirdos you're likely to meet at this hour.'

'I'll take my chances.'

'You're being stupid.'

She knew she was as she slammed the door, and

walked away. Crossing The Meadows on her own at half past four in the morning was not a smart idea. Crossing The Meadows when the snow was falling in ever-increasing, large, whirling flakes was even crazier. Even if no one approached her she would still have all those streets to walk unless she could find a taxi, but she had to get away from him. She had to get away because she knew if she didn't she was going to burst into tears, and no way was she going to give him the satisfaction of knowing she cared so much, and so stupidly longed to be the one woman who might change him.

'Brontë, wait a minute!'

She heard the sound of his door closing, then a muttered oath which meant he'd tripped over something. Serve him right, she decided grimly, not slowing her stride by an inch. If he thought shouting at her some more would get him anywhere, he was dumber than a rock.

'Brontë.'

He had caught up with her, but, when he clasped her shoulder and tried to turn her to face him, she shrugged him off.

'Go away, Eli. Just…just go away!'

'But I want to talk to you,' he insisted, coming round in front of her, barring her way.

'Then you're going to be sadly disappointed,' she retorted, trying her hardest to sound indifferent, but unfortunately a wayward tear slid down her cheek.

'Oh, jeez, *don't*,' he said, staring at her in complete horror. 'Brontë, please, don't cry.'

'I'm not,' she replied, her voice betraying her. 'I just… I've simply got something in my eye.'

And desperately she wiped her nose with the back of her hand, and it was that small action which cut him to the bone. That completely uncalculated, almost childlike, defiant little action which stirred and touched something deep inside him, something he couldn't pinpoint or define.

'You're getting wet,' he said, reaching out and gently brushing the snowflakes from her hair and eyelashes. 'Please come back to the ambulance before you catch your death of cold.'

'Old wives' tale,' she replied, her voice thick. 'Germs cause colds, not getting wet.'

'Do you always have to have the last word?' he demanded, putting his hands on her shoulders in case she was thinking of walking away.

It was all he meant to do, just to keep her there,

to stop her from doing anything rash, but, as he gazed down into her tear-filled eyes, she suddenly sniffed, and wiped her nose again, and those two actions were his undoing. Before he could think, before he could even rationalise what he was about to do, he bent his head and kissed her.

He meant it only to be a light kiss, a gentle kiss, but, as her lips opened under his, and he tasted her sweetness, her softness, and heard her give a small, shuddering sigh, that light kiss wasn't enough. Before he could stop himself, he had wrapped his arms around her, bringing her closer, closer, intensifying the kiss, deepening it, until all he was aware of was the fire and heat within him, and the need never to let her go.

And Brontë felt it, too, the same fire, the same heat, the same need, and she wound her arms round his neck, tasting him, his warmth, wanting so much to touch him, to feel his skin, but their bulky high-visibility jackets prevented that touch.

I want him, she thought, as she threaded her fingers through his hair to bring him closer, returning his kisses with a depth to match his own. I have always wanted him right from the first moment I saw him in Wendy's hallway when he walked right past me, but I can't let this continue because he'll leave me. He'll leave me just as he's left every other woman he's kissed, and it was that certainty which gave her the strength to jerk herself out of his arms, and back away from him.

'Don't, Eli, *don't*,' she said, her voice breaking on a sob. 'Don't flirt with me, play with me, lead me to believe you really care, because I can't stand it.'

'Brontë—'

'I'm not like you,' she said blindly. 'I can't live the way you do—being with one person one month, then another the next—and if I let myself get too close to you, you'll break my heart, and I don't think I'll ever be able to put the pieces back together again.'

'I would never break your heart,' he protested, his breathing harsh and erratic in the silence, and he reached for her again, only to see her back away still further.

'You might not mean to, you might not intend to, but you *will*.'

'Brontë—'

'Answer me one question,' she interrupted.

'How long would I get with you before you move on to someone else?'

He thrust his hands through his wet hair, his face impatient.

'How can I answer that?' he exclaimed. 'We'd just take it one day at a time, like every other couple do.'

'But we wouldn't be like every other couple, would we?' she cried. 'Because you've already told me you never want to settle down with just one woman, and I...' She took a long, juddering breath. 'I'm thirty-five years old, Eli, and I don't want what you call "fun." I want someone who will be there for me, someone who will care for me, someone...' Her voice broke again. 'Someone who will *love* me, and that's not you, is it?'

She saw indecision, and uncertainty, war with one another on his face, and then he shook his head.

'I'm sorry,' he murmured.

'So am I,' she said but, as she turned to go, he put his hand out to stop her.

'I won't let you walk back through The Meadows alone. If...' He bit his lip. 'If you can't bear to sit beside me in the ambulance I'll walk back to the station instead.'

'You could get mugged just as easily as me,' she pointed out, and he forced a smile.

'I'm bigger than you, tougher.'

'Are you?' she said, then shook her head. 'We'll go back together, but I want you to promise me something. I want you to promise there'll be no more flirting, or flattery, and definitely no more kisses, because I can't take any more of this. I really can't.'

He didn't say anything. He simply nodded and, as she walked back to the ambulance, not looking at him, not saying anything, all she could think was it didn't matter if he kept his promise because it was too late. She was already in love with him, and her heart was already breaking.

# CHAPTER SIX

*Sunday, 12:15 a.m.*

SHE'D always hated Saturday nights at the Waverley, Brontë remembered as she drove over the North Bridge. Saturday nights in A and E meant drunks, and RTCs, chaos and mayhem. They meant exactly the same thing for the ambulance service. From the minute they'd clocked on she and Eli had been working flat out and for that, at least, Brontë was grateful. Working flat out meant they hardly had time to talk to each other. Working flat out meant no awkward silences, or uncomfortable pauses.

Oh, who was she kidding? she thought, as she risked a quick glance at Eli. She'd expected tonight to be difficult, but what she hadn't expected was for Eli to be so angry. Not an obvious anger, not a blatant anger, but a simmering, bubbling, undercurrent of anger she could feel, almost taste, and normally she would have moved into what her brother, Byron, called her 'make everyone happy' mode. She would have talked and talked until she was sick to death of the sound of her own voice, but tonight she didn't.

Tonight she was tired, and weary. Tired from tossing and turning sleeplessly, and weary from alternatively telling herself she'd made the right decision, and then berating herself for being an idiot not to have simply grabbed those two months from Eli and enjoyed them while she could.

'Left, you should have taken a left there for Bread Street,' Eli declared as she turned right at the junction.

'Are you sure?' she replied. 'I thought it would be faster if I cut along the Grassmarket?'

'It's not.'

'Right,' she muttered, and began to reverse back up the street.

'What on earth are you doing?' Eli exclaimed.

'Going back,' she protested. 'You said, turn left—'

'That doesn't mean I expected you to reverse back up a one-way street!'

Damn, but she hadn't even seen the one-way warning sign, and swiftly she drove back down the road, crunching the gears so badly even she winced.

'Sorry,' she mumbled.

'You do realise this detour means not only is our patient still waiting, we also can't possibly hit our target arrival time?'

'I *said* I was sorry,' she retorted. 'I made a mistake, okay, and I'm *sorry*.'

Just tonight and tomorrow night, she told herself as she drove on and heard Eli mutter something which didn't sound like, 'Apology accepted.' I've only got to get through tonight, and tomorrow night, and then I'll never have to see him again, but how was she going to get through those two shifts when she doubted if she could stand even another hour in his company?

'Watch the corners,' Eli declared as she took one too fast. 'There's at least a foot of snow out there now, and it's fallen on ice.'

'Look, would you prefer to drive instead of me?' she snapped.

'Perhaps I should, if you're going to be silly,' he replied tersely, and she gripped the steering wheel until her knuckles showed white to stop herself from decking him.

Silly. He thought she was silly. Well, maybe she

was, to have fallen in love with a man who quite patently didn't give a damn about her unless it was to add another notch to his bedpost. If anyone had the right to be angry, it was her. His heart wasn't broken. He would go on and find someone else fast enough, whereas she... She swallowed painfully as she cut across town, then came down Bread Street. Don't think, she told herself, don't remember the touch of his lips, his hands, because that's a fool's game.

'You've just gone straight past the house,' Eli declared. 'It's number 22, Bread Street.'

She knew it was. The caller had said, 'Harry Wallace, twenty-nine years old, in an apparently catatonic state, his mother with him, number 22, Bread Street,' and yet she'd gone right past the house without even noticing.

*Pull yourself together,* her mind whispered as she turned the ambulance too fast and swore as she felt the wheels spin slightly on the compacted snow. *This patient needs you, so pull yourself together.*

And Eli stared grimly out of the window, and was tempted to ask Brontë what her problem was, except he already knew the answer. According to her, he was the problem, and that accusation only made him all the angrier.

If she had simply turned down his dinner invitation this morning, he could have lived with it. Okay, he would have been irritated, annoyed, because women didn't normally turn down the chance to become involved with him, but she hadn't simply turned him down. She'd said she wouldn't go out with him because he had broken too many hearts which was nonsense. Okay, so Zoe had created the mother and father of all scenes when he'd left her, but that was because he'd made a mistake, not because he'd left a trail of heartbroken women in his wake.

'Will I take the defibrillator, just in case?' Brontë asked as she pulled the ambulance to a halt, and opened her door.

'You should know by now we always take everything,' he replied acidly.

She opened her mouth, then clamped her lips shut, and he had to bite down hard on the angry words that sprang to his lips, too, as she retrieved the defibrillator, then stomped up the path to the house. The sooner she was out of his life, the better. The sooner she left the station, the quicker he would forget her, and her troubling accusations.

*And her kiss?* his heart whispered. *Will you forget that, too?*

It was just a kiss, he told himself. No different from any other woman's kiss, but no matter how often he had told himself that since this morning he knew it wasn't true. It had been a kiss like no other kiss. A kiss that had made him feel as though he had somehow—oddly—finally come home, and he didn't want to feel like that, or even think it.

'I'm so glad you're here,' Mrs Wallace said when she answered Brontë's knock. 'My son... Harry...he suffers from bipolar depression. I don't know whether he's been deliberately not taking his medication, forgotten to take it, or if this is something new, but I came round when I couldn't get an answer to any of my phone calls, and...' She spread her arms helplessly. 'He's just sitting there.'

Harry Wallace was, and though both Eli and Brontë attempted to get some response from him, Harry either wouldn't, or couldn't, communicate.

'Is this the only medication your son is taking,

Mrs Wallace?' Eli asked, lifting a bottle of pills from the mantelpiece, then putting it back down again.

'I'm afraid I don't know,' Harry's mother replied. 'He was diagnosed with bipolar almost ten years ago, and he's been on so many different pills since then. His bathroom cabinet's full of bottles, some with just a few pills in them, some completely full.'

'Show me,' Eli said and, as he followed Mrs Wallace out of the room, Brontë chewed her lip.

Discovering what pills Harry Wallace had in his bathroom wouldn't be hugely helpful. Knowing what he was currently taking, however, would, and, quickly, she crossed to the mantelpiece, and lifted the bottle of pills. As she'd hoped, they had a GP's name on them and she pulled her mobile phone out of her pocket.

That the GP was not happy about being called at one o'clock in the morning was plain, but he eventually gave her a list of Harry Wallace's current medications, and she had just finished writing the names down on a scrap of paper when Eli and Mrs Wallace returned to the sitting room.

*Who are you talking to?* Eli mouthed at her,

and she turned the piece of paper over, scribbled the word GP on it, and held it up to him. To her complete bewilderment he began making slicing motions across his throat, clearly wanting her to end the call, but she had no intention of being so rude, not when she'd just asked the GP whether he could come to Bread Street.

'Brontë, *end the call.*'

She could hear the barely suppressed anger in Eli's voice, and she gritted her teeth. It was one thing to be angry with her over something personal, but when he brought that personal antagonism into their professional lives…

'Thanks for your help, Dr Simpson,' she said into her mobile phone, flipped it shut and turned to face Eli. 'Harry's GP is coming, and I have a list of the medications he should be taking.'

Eli took the piece of paper she was holding out to him, then smiled reassuringly at Mrs Wallace.

'As we're not sure whether Harry has simply missed a dose, perhaps inadvertently taken more than he should, or the medication is no longer controlling his bipolar, I think it would be best if we take him to hospital, and let the experts examine him.'

'But, Eli, Dr Simpson is on his way here,' Brontë declared pointedly, but she might just as well have been talking to the wall because he was already helping Harry Wallace to his feet.

'If you'd like to come with us, Mrs Wallace,' Eli said, 'you're more than welcome.'

But I'm clearly not, Brontë thought, as Mrs Wallace reached for her coat and Eli ushered Harry out into the hall without even a backward glance at her.

What in the world was he doing? Okay, so he was clearly still angry with her about what had happened in The Meadows last night, but Dr Simpson was going to arrive and find Harry Wallace's house in darkness. Well, it wasn't good enough, she decided, and she was going to tell Eli that in no uncertain terms after Harry had been safely admitted to the Pentland.

It took longer than she'd thought. Harry Wallace seemed to be quite well known to the staff of A and E, and to Men's Medical, and it was more than forty minutes before they could finally get away, but the waiting didn't decrease Brontë's anger. If anything, it fuelled it because Eli seemed to be simmering, too, and when she drove away from the hospital, she didn't drive far. Just round the corner, then she pulled the ambulance to a halt.

'Okay, I want an explanation, and it had better be a good one!' she exclaimed. 'It's one thing to be angry with me personally, but when it affects a patient's treatment—'

'Brontë—'

'You're obviously in a major strop because I got a list of the medications Harry Wallace is currently taking,' she interrupted, not even bothering to hide her fury. 'All right, so I thought of phoning his GP, and you didn't, but what the heck does that matter? I know you're the accredited paramedic, and I'm not, but surely a good idea is still a good idea?'

'Brontë—'

'And I persuaded Dr Simpson to come round to the house,' she continued, on a roll now. 'Dr Simpson, who, I might add, is going to have a completely wasted journey because you simply bundled Harry into the ambulance and took him to hospital!'

'How do you know Dr Simpson is coming?' Eli said with a calmness which was infuriating.

'Because he *told* me he would!' she retorted. 'Did you want to talk to him yourself—is that what this is all about? You're piqued because the number cruncher took the phone call?'

'I'll ignore that,' he replied, his tone considerably harder than hers, 'but I will repeat, what proof do you have Dr Simpson is on his way?'

She stared at him blankly. 'What are you talking about? I phoned him—'

'Brontë, we *never* call a GP from a landline, or a mobile. If we need to speak to a GP, or a social worker, or a CPN, we ask EMDC to call them for us because then all the telephone conversations are recorded, which means if someone says they are going to attend they'd better.'

'But—'

'There are some wonderful GPs out there, some terrific social workers,' Eli continued, 'but there's a small minority who are more than happy to shunt all their responsibility onto the ambulance services rather than getting off their butts and doing what they're paid for. With no way of proving your phone call, we could be sitting in that house until doomsday waiting for your Dr Simpson, and he could deny he'd ever made that promise.'

Eli was right, Brontë realised with dismay as

she stared at him, but knowing it didn't make her feel any better.

'Why didn't you tell me this before?' she demanded. 'If you'd just told me, kept me in the loop—'

'How was I to know you were going to do something stupid?'

Stupid. He thought she was stupid. He thought she was stupid, and silly, and a doormat, and the hurt she had felt ever since he'd thrown that last epithet at her, combined with the pain she still felt from realising all he had wanted from her was 'fun', brought tears to her throat. Tears she was never going to let him see.

'Okay. Fine,' she said with difficulty. 'You know everything, and I'm the village idiot, so in future I'll simply drive your ambulance, and the only words I'll speak will be, "Where to?"'

'And now you really *are* being stupid,' he threw back at her.

She didn't say a word. She simply started the engine, and drove off down the road, but she didn't get far before their MDT began flashing a message.

'Suspected purple in the Potterrow. Police officer in attendance. Young boy who doesn't appear to be breathing.'

Brontë's eyes shot to Eli's. A 'purple' was ambulance code-speak for someone who was dead, but that wasn't the word on the screen which had caused her to suck in her breath sharply. It was the word *boy*.

Please, don't let it be him, she prayed, as she shot off down the road, heedless of the icy, snowy conditions. Please let it be somebody else, anyone else. She knew Eli was thinking the same, could see it in the tense way he was sitting next to her. He had tried so hard to help John Smith, and if the young boy was him…

'I'm afraid there's nothing you can do here, folks,' the police officer said when she and Eli got out of the ambulance. 'Judging by how cold he is I'd say he's been dead for a couple of hours. No sign of foul play that I can see so my guess is hyperthermia, though on these streets it could well be drug-related even though he's just a kid.'

A kid who was wearing a pair of threadbare trainers, thin denim trousers, and a tattered wine-coloured jacket, Brontë realised as she walked slowly towards the small figure lying huddled in the shop doorway.

Why did it have to be him? she wondered as she knelt down beside John, and took his cold, stiff hand in hers, automatically feeling for a pulse, although she knew there wouldn't be one. He had been so frightened of death, so afraid someone would kill him, and it hadn't been a person who had killed him. It had been the elements, and a society that had walked past him, ignoring his plight.

'Any pulse?' Eli asked, his voice gruff, and she shook her head.

'No, nothing,' she said, through a too-tight throat. 'Eli…'

He wasn't listening to her. He was already reaching for John and, when he'd lifted him up into his arms, he walked determinedly towards the ambulance, leaving her where she was, kneeling in the snow.

'If this weather doesn't change soon, I'm afraid we're likely to see more cases like this,' the policeman said sadly. 'And when it's a kid… It always hits hardest.'

'It hits hard no matter who it is,' Brontë replied, thinking of Peg, and the others at Greyfriars. 'And he wasn't just a kid,' she added. 'He had a name, and his name was John. John Smith.'

Wearily, she went to the ambulance but, when she opened the back door, what she saw there stopped her in her tracks. Eli was performing CPR on John. Desperately, and frantically, he was performing CPR.

'Eli…' She bit her lip. 'It's no use. He's been dead for at least two hours.'

'People can survive for longer than that if they're in a state of suspended animation,' he muttered. 'There are well-documented cases of people being pulled out of freezing rivers, and they've been brought back.'

But John hasn't been in a river, she wanted to point out, but she didn't. Instead, she climbed into the back of the ambulance, closed the door, and began affixing an Ambu bag.

'Epinephrine, and the defibrillator, Brontë,' Eli declared.

Obediently, she did as he asked, though all her professional knowledge told her neither things would help.

'Set the power to two hundred,' he ordered.

She did, and then she stepped back from the trolley and closed her eyes. She didn't want to watch this, couldn't bear to watch this. No power on earth could bring John back, and to see Eli's stricken face, to watch him frantically do everything he knew, try everything he could, required more courage than she possessed.

Three hundred, three hundred and sixty joules… Numbly she upped the power every time Eli asked her to but, eventually, she knew she had to say something and when, after twenty minutes, Eli reached for the paddles again, she put her hand on his arm to stop him.

'He's gone, Eli,' she whispered, her voice breaking. 'We have to accept he's gone, and nothing we do is going to bring him back.'

'I can't let him die, Brontë,' he replied, anguish thickening his voice. 'Maybe if I just keep on trying…if we keep on shocking him…'

'He's *dead*, Eli,' she said, her voice suspended. 'We're too late. You have to accept we're too late.'

For a second, she thought he was going to argue with her, continue with his attempts, then she saw his face twist and, when he sat down with his head in his hands, she gently kissed John's cold forehead.

'Be at peace now,' she murmured, fighting to contain her tears. 'Be at peace, and if there is a heaven be happy there.'

Impotently, she brushed the remaining flakes of snow from his hair. He looked so young, even younger than the fourteen years she'd guessed him to be, and, though she didn't want to do it, she carefully pulled a sheet up over his face, then turned slowly to Eli.

'I always say we can't win them all,' Eli said, his voice muffled. 'That sometimes the only victory we get in this job is if we can manage to keep somebody alive long enough to get them to hospital, and their families can arrive and say goodbye to them, but who was there to say goodbye to John, Brontë? *Who?*'

Instantly, she knelt down in front of him and tentatively covered his hands with her own.

'We were,' she replied. 'We were here, and though we didn't get here in time, I'm sure he knows you tried. You tried so very hard.'

'He ought to have his whole life ahead of him,' Eli exclaimed, raising his eyes to hers, eyes that were full of misery and pain. 'He ought to have a future, a home, a job, and now... Why, Brontë, *why?*'

'I don't know,' she declared. 'I don't know why some people open the wrong door, take the wrong corner.' She glanced over her shoulder at the small, sheet-wrapped figure. 'We'll have to take him to A and E, won't we, so his death can be formally registered?'

Eli nodded. 'And then his details will go to the procurator fiscal who will arrange for an autopsy.'

'Does there have to be one?' she protested. 'He's already been through so much.'

'I'm afraid it's the law. If someone dies unexpectedly, there's always an inquest.' Eli glanced towards the front of their cab, and let out a bitter laugh. 'Guess what, Brontë? Your bosses will be really pleased with ED7 tonight. You made that call in under seven minutes so you can notch John up as one of ED7's success stories even though he's dead.'

*'Don't,'* she begged, hearing the raw pain in his voice. 'Please, *don't*. I know how you're feeling, and this isn't how I wanted his life to end either. Eli—'

He slipped his hands out from under hers, and got to his feet abruptly. 'We'd better get going. There'll be other cases—people who need us.'

'Yes, but...' She scanned his face. 'Eli, I think after we've taken John to the Pentland we should log out, take the rest of the shift off to decompress.'

'I'm fine.'

'You're not,' she insisted. 'I know I'm not, and it's EMDC regulations that when you've had a very difficult job you should come off the road.'

Anger flashed across his face.

'If I say I'm fine, Brontë, then I'm *fine*.'

He wasn't—she knew he wasn't, and neither was she—but she had no opportunity to argue with him. They had scarcely returned to their ambulance after they'd taken John to the Pentland when their MDT flashed up a message.

'Code red. Two-month-old child. Not waking up. Number 108, Nicolson Street.'

'Oh, damn,' Eli muttered, and Brontë felt the same as she put the siren on, and her foot on the accelerator despite the icy, snow-covered road.

It sounded very much like a case of sudden infant death syndrome. Every medic's worst nightmare, every family's worse fear. She hadn't thought this shift could possibly get worse, but it just had, and it got even worse when she reached Nicolson Street, and saw cars parked on either side of the road.

'I know,' she said as Eli opened his mouth. 'This is an emergency, so just park in the middle of the street.'

She did but, even as Eli pulled a medi-bag out of the ambulance, she could hear the sound of people crying from inside number 108. Crying was never a good sign. Crying meant they were probably too late, and they were. One glance at the baby was enough to tell Brontë the child had been dead for quite some time.

'She was all right when I put her to bed,' the young mother sobbed. 'She'd had her bottle, and she was fine—a little snuffly, but nothing else. She was fine, she was fine—and now…'

'I work nights,' her husband declared, rubbing a hand across his tear-stained cheeks, a muscle in his jaw twitching, 'and I just looked in, like I always do when I get home, and Jenny… She was just lying there, and I knew right away there was something wrong.'

Brontë looked helplessly across at Eli, but he was already scooping the baby into his arms.

'Hospital. This little one needs the hospital,' he declared, striding towards the door, and the child's distraught parents grabbed their coats and followed him. 'Blue us in, Brontë.'

Obediently, she hurried out of the house, but she didn't get far. A middle-aged woman was standing beside their ambulance looking distinctly annoyed.

'Look, are you going to be much longer here?' the woman declared. 'I have to get to my night shift at the supermarket, and I can't get past.'

Brontë stared at the woman in stunned disbelief, but Eli wasn't similarly reduced to silence.

'Madam, we have a critical case here,' he replied, 'and it will take as long as it takes.'

'Can't you be a bit more specific, time-wise?' the woman protested, and Eli drew himself up to his full six feet two with an expression on his dark face that would have had Brontë backing away fast.

'Madam, do you have children?' he declared tightly.

'Well, they're teenagers now—'

'Then I'd like to know how you would feel if I couldn't park outside *your* door because some insensitive, uncaring individual felt I was blocking her way!'

The woman reddened, but she wasn't crushed.

'I want your name!' she exclaimed. 'I demand

to know your name so I can make an official complaint!'

'If you can't read the name tag on my uniform, then I'm certainly not going to spell it out for you,' Eli retorted as he got into the back of the ambulance with the baby's parents.

'I'm going to report you—both of you,' the woman yelled after them as Brontë drove away.

Brontë fervently hoped she would. She would enjoy contesting that complaint but, as she drove to the hospital, she would have preferred, even more, to have been anywhere but where she was.

It was one thing to be in A and E when a SIDS baby was brought in, and quite another to drive through the dark Edinburgh Streets with that SIDS baby in the back of her ambulance. She tried not to look in her mirror because she didn't want to see Eli performing CPR on a baby who would never laugh or cry ever again. She tried to stop her ears to the sound of the heartfelt, wrenching sobs from the baby's parents, but she couldn't do that either. All she could do was bite down hard on her lip, and pray she would get to the Pentland fast.

'That was the worst journey of my life,' she

mumbled when the baby and its parents had been handed over to the care of the A and E staff.

'I know the regulations say I should have declared the child dead immediately,' Eli replied as he and Brontë left the waiting room, 'but those poor parents...' He shook his head. 'They need to know everything that could be done was done, and in hospital there are people who can help them rather than us just simply driving away, leaving them alone with their baby and their grief.'

He looked as upset as she felt, and she half stretched out her hand to him, only to withdraw it quickly.

'The woman in the street—the one who was complaining,' she said angrily. 'How can people do that, behave like that?'

'It doesn't happen often,' Eli declared. 'In fact, I've watched people park on roundabouts in an attempt to let my ambulance through. I've seen people trying to get themselves into almost impossible spaces so they won't hold me back. I've even scraped the sides of vans as I've tried to get to the hospital as fast as I can, and the drivers have just called after me, "Don't worry about it, mate, it's not important." The vast majority of the public are decent people, Brontë, but some...'

'But *why*?' she insisted. 'Why are some people like that?'

'Because those people live in a "me" world, Brontë. They truly don't care about anyone else, don't think about anyone else's feelings, just take what they want, and concentrate on themselves.'

And he was describing himself, Eli thought with sudden, appalled recognition. No matter how much he tried to deny it, no matter how much he didn't want it to be so, everything Brontë had said about him was true. He *was* arrogant, he *was* self-absorbed, he *was* blind to other people's feelings. How much damage had he inflicted on the women he'd had in his life without noticing it, or—even worse—caring? Okay, so he cared passionately about the homeless, about people who were down on their luck, people who were disadvantaged by circumstances, and society, but in his personal life…

All he'd ever thought about was himself, what he wanted, what made him happy, and he was horrified.

'Are you okay?'

He looked down to see Brontë staring up at him with concern.

'I'm fine—fine.'

'I really do think we should log off,' she said tentatively. 'You need time out—we both do—and—'

'Brontë, I've been doing this job a hell of a lot longer than you have,' he snapped. 'I don't need nannying, I don't need my hand held, so back off, will you?'

She wasn't going to, Brontë decided, as he strode away from her. She'd seen murder in Eli's eyes when the woman in the street had been complaining, and she knew that if she didn't forcibly take him off the road he was going to lose it completely.

'I could do with a coffee,' Eli muttered when Brontë joined him in their ambulance.

'Me, too,' she replied as she drove away from the hospital and turned right at the bottom of the road.

But not at Tony's. They would both have their coffees in their own homes because she was going back to the station and signing them both out whether he liked it or not. Yes, it would leave ED7 one ambulance short but, no matter what Eli said, neither he, nor she, could cope with anything else tonight.

Eli glanced out of the window, then back at her.

'This isn't the quickest way to Tony's.'

'I know.'

'Brontë—'

'We're not going to Tony's,' she continued determinedly, seeing the dawning realisation in his eyes. 'I'm taking us both off the road. No ifs, no buts, no argument,' she added as he swore long and volubly. 'We both need the rest of the shift off.'

'But—'

'My decision, Eli, my call,' she interrupted.

She could feel him fuming beside her, could sense his anger and resentment, but she wasn't going to back down, not even when they got back to ED7, and he got out and slammed the ambulance door with a look at her that would have killed.

'I'll see you tomorrow, then,' she called after him as he began walking away from her, but he didn't reply.

He just kept on walking, and she stared indecisively after him. She'd never seen him looking so low, so down, and part of her wanted to go after him, to say she felt the same way, she understood, and she took a step forward, only to stop.

*Distance, Brontë,* her mind warned, *you were going to keep your distance, remember?*

But he's so upset, she argued back, and what harm would it do to ask if he'd like some company, to maybe share a meal with her, rather than them both going back to their empty flats alone?

*Mistake, Brontë, big mistake,* her mind insisted, and for a second she swithered and then before she had even realised she had made a decision she was running across the forecourt after him.

'You're inviting me to your place for a meal?' he said slowly when she caught up with him.

'I just thought... It's been such an awful night... maybe you'd like some company,' she said, feeling her cheeks beginning to heat up under his steady gaze. 'It wouldn't be anything fancy—just what's in the fridge—but the offer's there if...if you want it, that is.'

She thought he was going to refuse—he looked very much as though he intended to—then, to her surprise, he nodded.

'On two conditions,' he said. 'Number one, I drive you home, and number two, we make a short detour to Tony's to pick up some take-away spaghetti and meatballs, and then neither of us have to cook.'

A take-away spaghetti and meatballs sounded wonderful, and it smelled even better when she'd unwrapped it in her small kitchen.

'What do you want to drink?' she asked, taking two plates out of her kitchen cupboard. 'I have a half-bottle of red wine left in the fridge, coffee, tea…?'

'Coffee, as I'm driving,' he replied.

'Comfy seats, and slobbing out in the sitting room?' she said, switching on the kettle. 'Or hard seats with a table in the kitchen?'

'Comfy seats every time,' he declared, and when she led the way into the sitting room he smiled as he gazed around. 'Nice room.'

'The flat's rented, but the furniture's all mine,' she said, pulling a coffee table over. 'It's nothing special, or valuable, just bits and pieces I've picked up from second-hand shops, and car boot sales over the years.'

'You like old furniture?' he said, sitting down on the sofa.

'I like the idea of lots of people having owned something before me,' she said, unzipping her jacket and throwing it over one of the chairs. 'That they've polished something, touched it, maybe left a little bit of themselves, their hearts, in it.' She laughed a little uncertainly. 'And now you're thinking you're having dinner with a fruit cake.'

'A romantic,' he said firmly. 'I think you're a romantic.'

Did he mean that as a compliment, or a condemnation? She wasn't sure and nor was she about to ask.

'I'll see if the kettle's boiled,' she said instead. 'Don't wait for me—just start eating.'

He hadn't, she noticed when she returned to the sitting room. He'd taken off his jacket, but he was still sitting where she'd left him, lost in thought and, as she stood in the doorway, holding their coffees, a lump came to her throat. He looked so tired. Tired, and beaten, and she wanted to put down the coffees, and take him in her arms and say, 'I'll make everything all right,' but she couldn't. She couldn't make everything all right tonight, nobody could.

'Thought I told you to eat?' she said, all cheerleader bright, and he turned to face her with an effort.

'Sorry,' he replied. 'I was miles away.'

'Someone—a man who isn't always the smartest of men—once told me, Don't think,' she observed. 'And, on this occasion, I think he was right.'

Eli forced a smile. 'So, I'm a dummkopf now, too, am I, along with all my other faults?'

'Hey, why should I be the only one with a gazillion labels?' she protested, taking a seat opposite him, hating to see that forced smile on his lips. 'And you're not a dummkopf, just a very complicated man, and now eat something.'

'Yes, ma'am,' he replied, giving her a mock salute, and she managed to laugh, and picked up her knife and fork because she knew if she didn't she would put her arms round him anyway.

Because she didn't simply want to comfort him, she wanted him. Even though it would end in heartbreak, even though it would never be for ever, she knew she still wanted him, and she always would.

'Great meatballs,' she said, deliberately putting some in her mouth.

'Yup,' he said. 'I was wondering… Tomorrow's your last day at the station, so…'

'You can tell everyone at ED7 I'm giving the

station a glowing report,' she answered, and he shook his head.

'That's not what I meant. I was wondering what you were going to do, whether you were going to stick with this job, or...'

'I think we've pretty well established I'm rubbish at it.' She smiled. 'So when I hand in my report, I'll also be handing in my resignation.'

He sat back in his seat, his face concerned. 'Christmas is a lousy time to be out of work.'

'Maybe one of the big stores will be looking for a spare pixie, or an elf, for their Santa grotto,' she declared. 'Having said which, I'd have to lose a bit of weight first. Pixies and elves tend to be slender.'

'You look perfect to me.'

Her gaze met his, then she looked away fast. Oh, hell. Why did he have to have such intense blue eyes? Why couldn't he have had ordinary eyes, eyes which didn't make her heart jump around so much in her chest?

'You've just broken one of your promises,' she said, mock stern as she determinedly took another fork of spaghetti. 'The no-flattery one.'

'I wasn't flattering you.'

Oh, double hell. Change the subject, Brontë, she

told herself, and change it fast, but she couldn't think of a single thing to say, and he raised an eyebrow.

'Cat got your tongue?'

'Nope,' she replied. 'Just enjoying the food. As my mother used to say—'

She bit off the rest of what she'd been about to say. Mothers were not a good topic, not with Eli, and he clearly sensed her discomfort because he smiled a little wryly.

'What you said—about going to an agency to try to find my mother,' he began, and she cut across him fast.

'I'm sorry, I should never have suggested that. I can fully understand why you wouldn't want to go down that road—'

'I already have. I went to an agency two years ago because I thought…' He sighed. 'I don't know what I thought. Maybe I was looking for closure like you said, or maybe I just wanted to ask her, face-to-face, why she left me.'

'Did…did you find her?' she said hesitantly, not really wanting to know, but knowing she had to ask.

'I was too late. She… She died six months before I started my enquiries.'

Nothing could have prepared her for him revealing that, and nothing could have prepared her for the bleakness she could see in his face.

'Oh, Eli, I am sorry,' she said softly, 'so very, very sorry.'

'She took her reasons for leaving me to the grave, so now I'll never know why she did it,' he murmured, his eyes dark. 'Maybe if I was a "think the best of people" person like you, I could pretend. I could create a scenario, a perfect resolution, but I'm not like you.'

'You don't have to be,' she said, hating to see the pain in his eyes, wanting so much to ease it, but not knowing how, 'but I do think you need to let it go, or it will never stop hurting you.'

'Yeah, well… Sorry,' he added, 'I've really put a dampener on our dinner, haven't I?'

'I'm just flattered you felt you could tell me,' she said, and saw one corner of his mouth turn up in an attempt at a smile which didn't deceive her for a second.

'I thought you said flattery was a no-no?' he pointed out.

'Oh, very funny,' she said, 'now eat before all this lovely sauce gets cold.'

He did, and she did, too, but as they ate she

knew she could have been eating anything for all the impression it was making because all she was aware of was him. Him sitting in her sitting room, him just an arm's stretch away from her. Every time he pushed his hair back from his forehead, she thought, *Let me do that, I want to do that.* Every time he moved in his seat she tensed, hopeful, expectant.

Stop it, Brontë, she told herself, *stop it*. He's just a man, just a very handsome man, but he wasn't just a man, and she knew he wasn't.

'Something wrong?' he asked, looking up and catching her gaze on him.

'No—absolutely not,' she said brightly, much too brightly, and, because she was nervous, she forked up too much spaghetti only to watch in dismay as some of it dropped off her fork and landed with a soft splat on the front of her shirt.

'You are a klutz, aren't you?' Eli chuckled. 'No, don't stand up,' he added as she made to do just that, 'you'll get it on the carpet.'

And before she realised what he was going to do, he had picked up his napkin, and begun wiping the sauce and spaghetti from her shirt.

And it was torture. The most exquisite form of torture as he wiped down the front of her shirt in a smooth, rhythmic movement, and she felt his fingers through the napkin, through the material of her shirt, hot on her breast, and she knew when he caught his breath because hers caught at exactly the same moment.

'I think...' She heard him swallow. 'I think you're respectable again.'

Slowly she raised her eyes to his.

'Am I?'

Even to her own ears her voice sounded husky, and she saw him crumple the napkin into a tight ball.

'I shouldn't have done that, and I think...I think I should go now,' he said.

She didn't want him to go. She didn't want him to leave. She wanted him, and it didn't matter to her any more if she could only have this one night with him. It didn't even matter if he whispered words he didn't mean, said exactly the same things to her that he'd said to dozens of other women. She had no pride left. She just wanted him.

'You don't have to go,' she said, saw his pupils darken, then he got to his feet fast, shaking his head.

'Brontë, I do. You want more than I can

give—you know you do—and you deserve more.'

She stood, too, and, before he could evade her, she put her hands on his chest, and felt his rapid heartbeat.

'Right now, I want you,' she said softly. 'No strings, no promises, I just want you.'

'Brontë—'

She silenced him by standing up on her toes, and kissing him, with no reserve, no holding back, exactly as she'd kissed him in the snow what seemed—oh—like a lifetime ago now, felt him hesitate for a heartbeat, and then he was holding her tight, and kissing her back with a desperate, urgent need.

Heat, all she could feel was a pulsing, throbbing heat, and when he pulled her shirt free from her combat trousers, and slid his hand up underneath it to cup and stroke her breast, she shuddered against him, wanting more, so much more.

'Brontë, think about this,' he groaned as her fingers fumbled with his shirt buttons while she planted a row of kisses along his collarbone.

'I have,' she said, pulling his shirt apart so she could look at him, could see his beauty, his strength. 'And I have never been more certain of anything in my life.'

And she wasn't, she thought, as she claimed his lips again, and felt him slide her shirt down off her shoulders, felt her bra go in an instant, and then she heard his sharp intake of breath.

'I could kill the man who did this to you,' he said, his voice tight, vicious, as he stared at the ugly scar on her chest, but, when her hands instinctively came up to cover herself, he caught them, and pressed a kiss into each palm. 'He hurt you. He made you afraid, and I don't want you ever to be hurt, or afraid, again.'

And you'll hurt me so much more when you go, she thought, but she didn't say those words.

'Just make love to me, Eli,' she whispered instead. 'That's all I want and need right now—just for you to make love to me.'

And somehow they made it to her bedroom, and soon their clothes were gone, and they were skin to skin, on her bed, his body hard and muscular against her, and his kisses weren't enough, his touch wasn't nearly enough.

But he was hesitating still, she knew it, sensed it, knew he was holding himself back, and she knew why, and she kissed him harder, with greater intensity, and reached down to stroke him, stroke him, until he gasped under her fingers, and it was then he slid into her, hot, and hard, balancing himself on his elbows, his eyes fixed on her.

And he said her name, and it sounded almost like an apology, but she didn't want an apology. She arched herself up against him, forcing him on, so that he slid even deeper into her, and she bit her lip as he began rocking into her, over and over again, and she could feel it coming, feel the blood surging to her fingertips, her toes, every part of her, knew she was reaching the precipice, and she wrapped her legs round him, drawing him further inside her, and then she broke. Broke and began spiralling and shaking beneath him, and he came, too, and as he did, he laid his head between her breasts and she heard him give a sigh that was almost a groan.

# CHAPTER SEVEN

*Sunday, 9:30 a.m.*

ELI lay on his back and smiled as he gazed down at Brontë. She was lying curled up against his side, her head resting against his chest, and gently he placed a kiss on top of her head, and tightened his hold on her.

Last night had been the most incredible night of his life. He'd made love to her twice, and each time it had been wonderful, but that second time… That second time had been special. Slower, less frantic, it had been like nothing he had ever experienced before. It had felt almost as though he had come home which was crazy because he'd never had a home, not a proper one, but Brontë had taken him somewhere he'd never been before, and it was somewhere he didn't want to leave, not ever.

She stirred in his arms, almost as though she had read his mind, then she stretched against him, her full breasts rubbing lightly against his ribcage, sending a tremor of arousal through him, and his smile widened as she opened her eyes.

'Morning, sleepyhead,' he said.

For a second, she looked confused, as though she wasn't a hundred per cent certain where she was, then her grey eyes softened for an instant, and then, just as quickly, she looked away, and suddenly—and unaccountably—he felt cold.

'What time is it?' she asked.

Was it his imagination, or did her voice sound carefully neutral? Unconsciously, he shook his head. Imagination, it must be his imagination. He knew he had pleased her last night. Hell, just thinking about the tiny cries, and sighs, she'd given was enough to turn him on all over again.

'Nine-thirty,' he replied. 'Which means we have a whole day ahead of us before work tonight. A whole day in which we can do whatever we want, and I've already thought of some pretty interesting things we can do. For example…'

Gently, he slid his hand up her side and began brushing his fingers across one of her nipples. It hardened instantly but, when he moved to the other breast, she eased herself out of his arms, and moved further away from him in the bed.

'I'm afraid I have plans for today,' she said. 'Things to do, places to go and a report on ED7 to write.'

'If you need to go shopping, I can carry your bags, and as for the report, I can help you there, too,' he said, reaching out to cup her chin only to see her turn her head away.

'You can't help me,' she replied. 'The things I need to do… Only I can do them.'

'Okay.' He nodded, regrouping quickly. 'How about I keep you supplied with coffees all day, then rustle you up a delicious lunch, followed by an equally jaw-dropping dinner? You have to eat, and if you've got someone like me who—' he grinned '—is not only good in the bedroom, but also pretty damn good when it comes to wielding a saucepan, why not use me?'

'Today's not a good day for me, Eli,' she muttered. 'I've made plans—plans I can't cancel.'

'Then what about tomorrow—or the day after?' he said, feeling a chill begin to creep around his heart. 'I'll be on a three-day break because I'll just have finished a seven-day block, and if you're handing in your resignation I can check out Santa grottoes with you to make sure you don't end up working with a load of licentious pixies.'

She didn't laugh—she didn't even smile—and she still wasn't looking at him, he noticed.

'Actually, I thought I might go and stay with my sister for the next few days,' she declared.

'But, Brontë, you don't even *like* your sister,' he replied, and she shrugged.

'Yes, well, she's still family.'

If it had been anyone else he would have said he was getting the brush-off, but this was Brontë. Brontë who didn't play games, didn't toy with other people's emotions, so the fault had to be his, and he sat up, tucking the duvet round her so she wouldn't get cold.

'Brontë, listen to me,' he said. 'I don't know what I did wrong last night—'

'You didn't do anything wrong,' she broke in. 'It was great, just great, but I'm leaving ED7 today—'

'Which doesn't mean we can't see each other any more,' he protested. 'It doesn't mean this has to end.'

He saw her fingers grip the sheet.

'Eli, I had a great time last night—fabulous, truly—but can't we just leave it at that?'

'Leave it at that?' he echoed. 'Brontë, in case you've forgotten, we made love last night, and I think we should at least talk about it. I've obviously upset you in some way—'

'You haven't—not in the least. Look, as you told me before,' she continued as he tried to interrupt, 'there's mediocre sex, okay sex and great sex, and though the sex last night was pretty good, my feeling is we should quit while we're ahead, shake hands, wish each other well, and move on.'

*They should shake hands and wish each other well? They'd had great sex, but they should quit while they were ahead?*

She was making what they'd shared sound so clinical, so unemotional, and it hadn't been like that, at least not for him, and he caught hold of her chin with his fingers, and forced her to look up at him.

'Brontë, *talk* to me. Don't give me all these platitudes, all this…this crap. *Talk* to me.'

'I would if there was anything left to say,' she replied, her eyes meeting his briefly, then skittering away. 'We had a great time, but can't you just accept that's all it was? Now, do you want to use the shower first? I don't want to hurry you—make you feel I'm throwing you out or anything—but…'

She was throwing him out. No ifs, no buts, she

was throwing him out, and he didn't want to be thrown out. He wanted to hold her. Not even to make love to her again, though his body would have welcomed that with unbridled enthusiasm; he just wanted to hold her, to try to recapture that wonderful feeling of belonging he'd experienced last night.

'Brontë—'

'So are you showering first? It's just, like I said, I have—'

'Things to do, people to see and a report to write,' he finished for her grimly. 'Fine. Absolutely fine.' He threw back the duvet, and stood. 'Do I get a cup of coffee before I go, or would that interrupt your schedule too much?'

'You know where the kitchen is,' Brontë replied.

And she turned on her side when she said it, so all he could see of her was her bare back which meant she was well and truly shutting him out.

Well, fine, he thought, angrily. If that's what she wanted, then that was just *fine*, and he walked out of the bedroom, and slammed the door and didn't see Brontë bury her face deep in her pillow so he wouldn't be able to see or hear her tears.

\* \* \*

'I don't see anyone,' Brontë declared when she turned into Richmond Street. '"Elderly woman lying in the street," Dispatch said, but I can't see anyone.'

'Drive to the top of the road, then come back down again,' Eli replied.

It was on the tip of her tongue to say, 'And, like, *duh*, so you didn't think I was going to think of that?' but she didn't.

Perky, and upbeat, Brontë, she told herself. You are going to be perky and upbeat for the whole of this shift if it kills you, and quickly she drove up Richmond Street, then back down again.

'Well, unless she's the invisible woman,' she declared, 'I'd say she's either just wandered off, or it was a hoax call.'

'Someone's waving to us from outside that house,' Eli said suddenly. 'Pull over.'

Obediently, Brontë did as he'd said, and an Indian gentleman walked gingerly over the snow towards them.

'Are you looking for the lady who fell?' he asked, and when Brontë nodded he looked a little awkward, a little guilty.

'She is in my house. I know you are not supposed to move someone who is hurt,' he added as

Eli sucked in his breath, 'but it is so cold out here, and my wife, Indira, she said we couldn't leave her, not in the snow, not when she was clearly in such pain. My name is Mr Shafi, by the way.'

'That was very kind of you, Mr Shafi,' Brontë said quickly before Eli could say what she knew he was itching to say. 'Can you tell us anything about her?' she continued, as she reached for a medi-bag.

'Her name is Violet Swanson,' he replied, ushering Brontë towards his home. 'The poor old lady... She had been to the chapel, to say her prayers, and she slipped on the snow on her way home. Indira...she thinks perhaps the lady's arm is broken.'

Violet Swanson's arm *was* broken, and she was clearly in considerable pain.

'So stupid,' she murmured. 'Such a stupid, stupid thing to have done. I know I should have waited until morning to go to church, but it's very peaceful there at night, very comforting.'

'You'll have to go to hospital,' Eli declared. 'I'm sorry, but you really must,' he added when Violet Swanson began to argue. 'That arm needs to be X-rayed, and put in plaster. Is there anyone you'd like us to contact, to say where we're taking you?'

'I'm a widow, dear, live on my own,' Violet Swanson replied, wincing sharply, as Brontë began to strap her arm across her chest to keep it secure. 'No family. Archie and I were never blessed, but then you don't always get what you want out of life, do you, so you just have to make the best of things.'

Which is what I'm going to do, Brontë thought, deliberately avoiding Eli's eyes. Last night... It had been the most wonderful night of her life, but she'd known it was just that. One night. One incredible night she would be able to look back on, and remember, and though she'd remember it with some pain—the pain of knowing he was the man she wanted, but could never have—it was better to walk away now than live with Eli for two months and then have him leave. That, she knew, she would never survive.

'Mrs Swanson, do you think you can walk?' she asked. 'We have a carry-chair—'

'Of course I can walk,' Violet protested, but from the way she swayed when she stood, it was patently obvious that not only was she in a lot more pain than she was admitting, her whole system had taken a severe shock.

'The carry-chair,' Eli said firmly.

'Thank you for helping,' Brontë told Mr Shafi and his wife, as she helped Mrs Swanson into the chair. 'It was very kind of you.'

The man shrugged. 'What else could we do? We could not leave the poor lady, turn our backs on her.'

A lot of people could, Brontë thought when they'd got Mrs Swanson safely settled in the back of the ambulance, and she set off for the Pentland. In fact, too many people did. Too many people walked on by, thinking, Someone else's problem, not mine.

'It's a pity there's not a lot more people in the world like that nice couple in Richmond Street,' she observed after they had delivered Mrs Swanson to A and E. 'People who are willing to put themselves out, to help others.'

'To be fair, some people prefer to keep themselves to themselves,' Eli replied. 'They don't want to let down their guard in case people encroach on their personal space.'

Like me, he thought, as he watched Brontë climb back into the ambulance. Always he had been the one who'd been in control in all of his relationships, the one who had decided when it was over, but this time...

For the first time in his life he was completely out of his depth. For the first time in his life he didn't know what to do. Maybe if he talked to Brontë again. Maybe if he waited until the end of this shift, then talked to her...

*And said what?* his mind asked.

That he didn't want her to walk right out of his life. That he wanted to spend time with her. A lot more time. That he'd miss her snippiness, and her laughter. He'd miss her silver-grey eyes, and her fringe which never would stay flat. He'd miss *her*.

But how could he get her back? His looks had always been his primary tool of seduction. His looks, and some well-practised flattery and flirtation, but none of that would get him anywhere with Brontë. He needed something else, and he would have to find it fast or she would walk right out of his life just as quickly as she'd walked into it.

'Code red. Male, aged twenty, collapsed and in pain. Number 82, Bristol Street,' the MDT screen read, and Brontë grimaced slightly.

Code red sounded ominous, but Eli hadn't said to switch on the siren. Actually, Eli hadn't said very much at all since they'd come on duty, but he was watching her, she knew he was. Watching her with a slightly puzzled expression in his eyes, as though he was trying to figure something out. She wished he wouldn't. She wished he would just take what she'd said this morning at face value, because it would be easier for both of them if he did.

*Will it?* her heart whispered as she drove down Melrose Street. *Will it really?*

It had to be, she told herself, because the alternative was so much worse.

'Turn left at the bottom of the road for Bristol Street,' Eli advised and, as she did, she blinked as she saw the row of elegant Georgian buildings.

'Whoa, but these houses are seriously stunning.' She gasped. 'I wouldn't be able to afford to live here even on a hundred times my income.'

'Money can't buy you happiness, or contentment,' Eli replied cryptically.

He was right. Number 82, Bristol Street might have been just as lavish inside as it was out, but its occupants weren't in the least to be envied.

'You the ambulance dudes?' a young man

asked, his voice distinctly slurred as he opened the front door.

'Yes, we're the ambulance dudes.' Eli sighed. 'Where's the casualty?'

The young man waved his hand vaguely. 'Dining room, I think. Or it could be the sitting room. Dunno, really.'

'Great,' Eli muttered, pushing past him. 'Let's hope somebody in this house is in a fit state to answer some questions.'

Brontë very much doubted it. That the son, or daughter, of the house had decided to throw a party was clear. She could hear the thumping sound of music and laughter, and there was discarded food, and empty bottles, scattered everywhere.

'I wonder where the parents are?' she asked as she followed Eli over the parquet entrance hall, feeling her boots stick with every step.

'Winter cruise, shooting in the highlands, down in London to take in a few exhibitions?' Eli suggested, and Brontë shook her head.

'You'd think they'd have more sense than to go off and leave a bunch of youngsters with no supervision. So, what do we do?' she continued. 'Search every room, or what?'

They didn't have to. A girl who couldn't have been more than eighteen appeared, looking scared and worried.

'Gavin's through here,' she said, pointing to a door at the end of the hallway. 'He's acting all funny—not making any sense at all.'

Brontë wasn't surprised when she saw the young man. He was lying on a sofa, curled up into a ball, clutching his stomach and moaning, but it was the rest of his appearance which told her immediately what the problem was. His pupils were almost pinpoints, his skin was clammy, his fingernails and lips had a bluish tinge, and he was twitching uncontrollably.

'What's he taken?' Eli said, as he crouched down beside the young man.

'A few beers—maybe a shot or two of whisky,' the girl replied, and Eli exhaled with irritation.

'Look, give me credit for some sense,' he declared as Brontë began strapping a blood pressure cuff to the young man's arm. 'Is it crystal meth, cocaine, Ecstasy, or heroin?'

'Gavin doesn't do drugs,' the young girl protested. 'He's just had a bit too much to drink, that's all.'

'Sweetheart, he clearly doesn't know where he

is, or even *who* he is, and you don't get muscle spasticity like that from booze,' Eli protested. 'So, I'm going to ask you again. What's he taken?'

'Eli, his BP is way too low, and his pulse is very weak,' Brontë murmured. 'We're going to have to tube him, and fast.'

'Did you hear that?' Eli demanded, and the young woman looked from Brontë to the young man in panic.

'I...I don't know anything,' she said.

'Look, what's your name?' Brontë asked, seeing Eli shake his head in despair.

'It's Joanna,' the young woman replied. 'My name's Joanna.'

'Joanna, Gavin is very sick indeed,' Brontë said, 'and, without wanting to frighten you, he could die. We need to know what he's taken so we know how to treat him,' she continued as the young girl let out a gasp, and tears filled her eyes. 'We're not here to make judgements, but to *help*.'

'My parents will kill me,' Joanna whispered. 'They don't know about this party, and when they find out...'

Out of the corner of her eye, Brontë could see Eli was beginning to insert an intubation tube into Gavin's trachea to help him breathe, but they needed to know what the young man had taken, and they needed to know it fast.

'Joanna, has Gavin had any seizures?' she asked.

'Seizures?' Joanna repeated blankly.

'Had a fit, thrashed about at all?' Brontë explained.

The girl nodded. 'He had one about half an hour ago.'

'Please, Joanna,' Brontë said softly. 'We need to know what Gavin's taken before it's too late.'

The girl bit her lip, then took a shuddering breath.

'Heroin,' she said, her voice thick. 'He's taken some heroin.'

'Do you know how much?' Brontë said, and the young woman shook her head.

'Brendan… He said it would make Gavin feel great. He's been feeling a bit down, you see. His parents have stopped his allowance because he crashed his dad's Mercedes, and it was just meant to pep him up a bit.'

Eli muttered something under his breath that was most definitely unprintable.

'How often has he taken heroin?' he demanded.

'I think this was the first time,' Joanna replied,

'but I don't know. I honestly and truly don't know.'

'Brontë, his BP is going down even more,' Eli said urgently. 'We have to go.'

'Can I come, too?' the girl asked. 'He's my sort of boyfriend, you see.'

A sort of boyfriend who might not survive the night, Brontë thought, as she helped Eli carry Gavin out to the ambulance, past youngsters who were still partying, apparently oblivious—or uncaring—about what was happening in front of them.

'Do you think he'll pull through?' Brontë asked after they had taken Gavin to the Pentland.

'We can but hope,' Eli replied, 'but the waste, Brontë, the damn waste of a young life, if he doesn't!'

'I know.' She sighed. 'All that money, all those advantages, and yet to be so stupid.'

'Yeah, well, I'm afraid all the money in the world doesn't buy you some plain, old-fashioned common sense.' Eli glanced down at his watch and sighed. 'And can you believe we've only been on duty for a couple of hours?'

Brontë couldn't. This shift felt endless, and

it was about to get worse, she thought when a message appeared on their MDT.

'Gang fight in Princes Street Gardens. Police in attendance. All available ambulances to attend the incident. Repeat, all available ambulances to attend, and offer assistance.'

Her stomach lurched. A gang fight. That meant a crowd, and she was being asked to drive there, to get out amongst it, to offer help.

'Brontë…'

She heard the instant concern in Eli's voice, the gentle understanding, and breathed in again, hoping it might work, but it didn't.

'I'm okay,' she lied, hating the betraying wobble in her voice as she switched on the ignition. 'I'll… I'll be okay.'

She wouldn't be, she knew she wouldn't. Already her heart was racing at just the thought of being amongst a crowd, and her palms were so sweaty she could barely grip the steering wheel.

*Failure,* her heart whispered, *you're a failure. People could be hurt out there, and you can't do anything to help them.*

She wondered what Eli was thinking. *Liability,* that's what he was probably thinking, that he was stuck with a liability. Well, at least she would get him there fast, she determined. At least he wouldn't be able to accuse her of deliberately dawdling, but it was easier said than done.

'Look, I know you don't want to go there—that crowds freak you out—but couldn't you drive just a little faster,' he said softly, and Brontë glanced at him helplessly.

'It isn't me. I've got my foot to the floor, but we're losing power, and I don't know why.'

'You did make sure she was filled up?' he said, only to smile a little ruefully when Brontë shot him a withering glance. 'Sorry. Stupid question. The trouble is she's old, clapped out.'

'What do you want me to do?' Brontë asked. 'We're not going to be much use ferrying casualties to hospital at this speed, and what if it breaks down on the way to the Pentland with a seriously hurt person in the back?'

Eli chewed his lip indecisively, then seemed to come to a decision.

'Get as close to Princes Street Gardens as you can. At least that way I can be of use, help the other paramedics.'

And they looked as though they could do with all the help they could get, Brontë thought when she reached the garden and saw not just an array of police cars, and ambulances, but a seething mass of fighting youths.

'Stay where you are,' Eli ordered as he pulled a medi-bag out from behind him, and opened his passenger door. 'When I get out, lock all the doors, call Dispatch and tell them we have a problem with the ambulance, and stay where you are.'

'But—'

'Keep the MDT display on, and if anything urgent comes in, try and attract one of the policemen's attention, but do *not* get out of the ambulance.'

'But, Eli—'

He was gone before she could say anything else and, for one brief moment, she saw him clearly, pushing his way through the fighting youths, head and shoulders taller than most of them, and then he was gone and she wrapped her arms around herself as a wave of nausea engulfed her.

Never had she felt so frightened. She could see the rival gangs quite clearly through her windscreen. Some were hurling stones, some were armed with bits of wood they'd clearly torn from garden fences, some were simply using their fists, and here and there, she saw the flash of metal in the moonlight. Knives. Some of them were armed with knives.

And the fight was moving closer to her now. Occasionally, somebody was thrown against the ambulance, and she could feel it shaking under the impact, could hear the thud of the body, see wild, enraged faces in front of her, and the noise… the noise…

She put her hands over her ears, and curled herself into a ball in the driver's seat, but it didn't help. She could still hear the oaths, the roars, the screams, and she wanted to put the ambulance into reverse, to get away, but even if the ambulance would go anywhere, Eli was out there, amongst all the mayhem, and he might need help, except she couldn't help him, she couldn't do anything but sit here, with her eyes tight shut, and wait.

A bottle crashing against her windscreen had her sitting bolt upright in panic. They couldn't get in, surely none of them could get in, and then—out of the corner of her eye—she saw something else. A young boy was lying beside one of the hedges. A young boy, with an ashen face, and closed eyes, who didn't look much older than John Smith had been. A young boy who had blood on his forehead, and blood dripping down from his fingers onto the snow below.

Desperately she scanned the melee in front of her. If she could just see a paramedic—any paramedic—maybe she could attract their attention, point out the boy, but all she could see was people fighting, The blood dripping from the boy's hand was forming a pool, and he was even whiter now than he'd been a minute ago. He was going to die. She knew as surely as she knew anything that he was going to die, all alone with no one to help him, and a sob broke from her.

Somewhere in the distance someone screamed, and she bit down hard on her lip until she tasted blood. If she'd been there when John had been dying she could have saved him, but she hadn't been there. She was here now, though, and that young boy lying so close to her needed her, and she could help him if she could only find the courage.

With shaking hands she dragged a medi-bag out from behind her, and took a shuddering breath. She could do this. She *had* to do this and, though her heart was pounding, she opened the ambulance door and got out.

\* \* \*

*Where the hell was she?* Eli wondered, as he scanned the now almost deserted gardens. He'd told her to stay in the ambulance, not to move, but perhaps she'd got so frightened she'd just taken off into the night which meant she could be anywhere.

'You've not seen Brontë O'Brian, have you?' Eli asked urgently as one of the paramedics from ED12 passed him. 'Small, golden-brown hair, big grey eyes? The government number cruncher.'

The paramedic shook his head.

'Mate, I doubt whether I would have recognised my own mother in that mob tonight. Bedlam. Sheer bedlam.' He glanced over his shoulder. 'Maybe one of the cops might have seen her.'

Eli nodded, and quickly hurried over to the small group of policemen who were standing by one of the railings, looking decidedly the worse for wear. One had a black eye, the other's uniform was torn, and one had a badly cut lip.

'I'm looking for my colleague,' Eli declared. 'Small woman, short golden-brown hair, big grey eyes. Have you seen her at all?'

'Can't say I have,' the policeman with the black eye replied. 'Maybe she went off in one of the ambulances?'

It was a possibility—a remote one—but a possibility, but as Eli turned to go, the policeman with the cut lip suddenly frowned.

'A small woman, you said?' he declared. 'With short, golden-brown hair?'

'That's her,' Eli said eagerly. 'Do you know where she is?'

'If it's the same woman you're talking about, I'm afraid she took a knife wound. Pretty serious, too, by the looks of it.'

Eli's heart clutched in his chest. A knife wound. Pretty serious.

No. No. *No*.

If she was badly hurt, if she was… No, he couldn't think that—wouldn't allow himself to think that—but he had to know, had to find out.

'Can you take me to the hospital?' he said.

'Sure thing, mate,' the policeman with the black eye replied. 'Which one?'

The Pentland, or the Waverley? Eli desperately tried to think, but his brain didn't seem to be working. All he could see was Brontë lying white, and cold, and bleeding, on the snow. Brontë slipping away from him on a sterile hospital trolley, without him, and him never being able to say what he knew now to be the truth. That the indefinable emotion he experienced every time he was near her, was love. The feelings of protectiveness, the need, and the rightness, he had felt last night, was love. It was an emotion he had told Brontë didn't exist. An emotion he had denied for the past thirty-four years, and he couldn't lose it now. Simply couldn't.

'Hey, mate, you okay?' one of the policemen asked as Eli dashed a hand across his eyes.

'Yes, I'm…I'm okay,' Eli replied, trying to blot out the image of Brontë lying on a cold mortuary slab. 'Can we go now?'

'No problem,' the policeman replied, but as Eli turned to follow him he saw a figure emerging from the gardens.

A figure, whose short pixie cut was sticking up all over the place. A figure who had a smear of blood on her cheek, and who looked exhausted, and before he even realised he was moving, Eli was running towards her, his arms outstretched, and when he reached her, he clasped her into his arms and enveloped her in a crushing embrace.

'Don't ever do that to me again,' he said, his voice breaking. 'Don't ever just disappear like that, not *ever*. When I got back to the ambulance and you weren't there…'

'I couldn't stay,' she mumbled into his chest. 'There was this young boy, and he reminded me so much of John, and I couldn't stay, just couldn't, and...' She raised her head to him, her eyes shining. 'I did it, Eli. I *did it*. I was terrified, and I thought I was going to faint—'

'Are you okay?' he interrupted, scanning her face. 'The blood on your cheek—is it yours?'

'No—no—it's someone else's,' she said dismissively, 'but did you hear me, Eli? I conquered my fear. After I'd treated the boy, I just kept going, and going, with other casualties, and I could do it, I could *do it*!'

'This past half-hour,' he murmured, not really listening to her, 'when I couldn't find you, and the policeman said someone looking like you had been taken to hospital—'

'That was Liz Logan from ED10. She got a nasty knife wound. I saw them take her away, but I think she's going to be okay because she was talking...' Brontë chuckled. 'Actually, she was swearing.'

Eli held her tighter. 'Do you have *any* idea how scared I was when I thought it was you?'

'You worry too much.'

'And you...' The relief he had felt at finding

her safe turned to anger as he remembered the agonies he'd been through, fearing the worst. 'What the hell did you think you were doing? I *told* you to stay in the cab. Didn't I tell you to stay there, but, no, Brontë O'Brian thinks she knows best, so out she gets, completely disobeying my order….'

She pulled herself out of his arms, and glared up at him.

'Order? You didn't give me any order. You just suggested I should stay in the ambulance, but, Eli, it was mayhem out there—'

'All the more reason for you to stay put.'

'—and people were hurt, bleeding. I couldn't ignore them, and keep my self-respect, and once I got outside, once I started treating people, I was fine, I was okay. I *am* fine, I *am* okay.'

'I'm *not*,' he said with feeling. 'Brontë, get in the ambulance.'

'There's no point,' she protested. 'It's not going anywhere until the breakdown truck arrives.'

'No, but we have an audience,' Eli declared, suddenly becoming aware that the three policemen were watching them with various grins on their faces. 'And I have more to say to you.'

'More to chew me out for, you mean.' Brontë

sighed. 'Don't you see what this means for me, Eli? It means I can be a nurse again, go back to doing what I love most. Or I could retrain, become a paramedic. Not at ED7, of course,' she added quickly. 'I wouldn't inflict myself on you—'

'I wouldn't mind.'

'But the big thing—the most important thing—is you said I'd get over my fear, and you were right, so, please…' She caught and held his gaze anxiously. 'Can't you be pleased for me?'

'I am pleased for you.' He nodded. 'But I still want you to get in the ambulance.'

'But—'

He wasn't taking any buts. He took her by the arm and steered her firmly towards the ambulance, and waited until she'd got in before he climbed into the passenger seat beside her.

'Look, I'm sorry you were worried,' she said immediately. 'I should have left a note or something—'

'Brontë…' He paused, and shook his head. 'You scared the living daylights out of me, but, believe me, what I'm going to say now is a hell of a lot more frightening.'

'Something's happened to one of the paramedics?' she said instantly. 'Liz Logan, she was hurt a lot more badly than we thought—'

'No, it's not Liz, or any of the other paramedics,' he interrupted. 'This...this is about me.'

'You,' she said with a frown. 'I'm sorry, but I don't understand.'

'I don't either,' he murmured with a wry and rueful smile, 'and, as I've never said this to anyone before, I'm probably going to mess it up big time.'

'Mess what up?' she declared in confusion. 'Eli, you're not making any sense.'

'I don't think I have since the first minute I met you,' he said slowly. 'You see...when I thought I might never see you again...when I thought I might have lost you for good... It was then I realised you're the most irritating, annoying, lippy, wonderful woman I've ever met, and I need you in my life.'

She blinked. 'Okay, I understood the annoying and the lippy part, but as for the rest...'

'I'm saying I love you, Brontë O'Brian.'

She didn't say anything. She simply stared at him blankly, and he felt his cheeks darken with colour. Surely he couldn't have got it so badly wrong? He thought—he was sure—she felt something for him, but maybe he'd been wrong.

'Aren't you going to say anything?' he said with an attempt at a smile that fooled neither of them.

'I'm waiting for the punchline,' she declared. 'What you just said... There's got to be a punchline.'

'No punchline.'

'But...' She shook her head. 'Eli, you've only known me for seven days, and for most of that time we've shouted at each other, and you don't believe in love. You told me you didn't.'

'I didn't until I met you,' he said awkwardly. 'I've never wanted any woman to stay with me forever, but I want you to.'

'Is this another pitch?' she said, and he saw her lip tremble. 'Eli, if this is another of your pitches, another of your lines, I swear I will never forgive you.'

Never had he regretted his reputation as much as he did now when he saw tears shimmering in her eyes. Never had he so much regretted all the women he had dated, and then so carelessly discarded.

'No—absolutely *not*!' he exclaimed, reaching

out to capture one of her hands in his. 'I wish I knew the words to say to convince you I'm telling you the truth. Brontë, I *love* you. I have never said that to a woman before, and I know I will never say it to any other woman. I love you, and I know I always will.'

'Is this about last night?' she said uncertainly. 'Because we slept together, you now feel guilty—'

'*No!*' He let go of her hand, and gripped her shoulders. 'Last night…last night you gave me something I've never had before. It wasn't simply sex,' he continued as she tried to interrupt. 'I've had enough sex to know the difference. It's not something I'm proud of, it's something I now bitterly regret—all the women I hurt without even realising I was hurting them—but I can't erase my past, can't change it.'

'Eli—'

'I told you once before you're one of a kind, Brontë O'Brian, and you are. Last night… Last night you made me feel whole, you made me feel complete, that I finally belonged, and I want you to stay with me for always, to marry me. If…' His eyes met hers. 'If you'll have me.'

For an answer, she reached out and caught his

face between her hands, her eyes large and luminous in the dark.

'You're smug, and you're arrogant, and you're a bit of a prat.'

'And a true bastard.' He nodded. 'Don't forget that one.'

'No, that one I won't allow,' she said, her voice shaking, 'but I love you, too, Elijah Munroe. Even when you called me silly, and stupid, and a doormat—'

'I didn't mean that,' he said, alarm appearing in his eyes. 'I was angry, just lashing out—'

'I know you were,' she said softly. 'Don't forget I've called you some pretty rough things when I was angry.'

'I just…' His voice broke. 'I don't want you to leave me, Brontë.'

She could see the shadows of his past in his eyes, and vowed that, even if it took her a lifetime, she was somehow going to erase them.

'Eli, I'm always going to love you even when you're crotchety, even when you get right up my nose, and even when we argue as we're bound to,' she said, her voice trembling slightly. 'I am never, *ever* going to stop loving you, and I am never going to leave you.'

And to prove it, she tilted her head, and kissed him, giving him everything she had, holding nothing back, feeling his familiar heat, his strength, as he wrapped his arms around her to kiss her back, and this time it was even better than before because this time she knew it was for keeps, and they would be together, always.

'So is that a yes, you'll marry me?' he said breathlessly into her hair when they had to draw apart to breathe.

'What do you think?' She chuckled.

'I want to hear you say it,' he insisted. 'I want you to say, "Elijah Munroe, I love you and I'll marry you as soon as we can."'

'Kiss me again, and then I'll give you an answer,' she replied, unable to prevent herself from teasing him just a little.

And he did kiss her again. Kissed her until she had to clutch onto his shirt, feeling breathless and giddy, so when their MDT bleeped insistently she groaned against Eli's mouth.

'I told them we were off the road,' she protested. 'I told them the ambulance had broken down and we couldn't take any calls.'

Eli glanced down at the screen, then his lips quirked.

'I think you'd better read what it says,' he declared.

'Why?' she said in confusion, and, when his smile widened she gazed down at the MDT, then across at him in confusion.

There, across the screen, were the words, 'We can't stand the suspense. Are you going to marry him, or not?'

'What the...? How can they...? How do they...?' She faltered, and then her cheeks flushed scarlet and she frantically stretched across the dashboard, and switched off the radio receiver. 'Oh, criminy, Eli, did you realise the radio was still on?'

He shook his head.

'You must have left it on by mistake when you patched EMDC to tell them our ambulance was out of commission,' he replied, reaching for her again only to see her put up her hands to fend him off.

'But that means everyone heard what you said, and everyone heard what I said,' she wailed. 'They were all *listening*, Eli.'

'I don't give a damn.'

'I do,' she protested. 'Oh, this is so embarrassing. Can't you see how embarrassing this is?'

'Only if you tell me you don't want to marry me.'

'But we're never going to live it down,' she declared, putting her hands to her hot cheeks. 'We must have sounded... Those kisses, do you think they heard those kisses?'

'I doubt they heard them.' He laughed. 'But I should imagine they managed to put two and two together during the long silences when we weren't talking, plus there was all that heavy breathing, of course.'

'It isn't *funny*, Eli,' she exclaimed, and he caught both of her hands in his tightly.

'Brontë, everyone heard me tell you I loved you. Everyone heard me ask you to marry me, so look on the bright side. I can't back out now even if I wanted to. Not if I don't want to be lynched.'

She stared up into his deep blue eyes uncertainly. 'And do you—want to back out?'

'Not now, not ever, so quit with the stalling, O'Brian. Will you marry me?'

There was love, and tenderness, and a great deal of uncertainty, in his face, and it was that uncertainty which tugged at her heart, and brought a tremulous smile to her lips.

'Of course I will, you idiot.'

And she leant towards him to kiss him again, but he put a hand out to stay her.

'Just one minute,' he said firmly, though his eyes, she noticed, were gleaming, and he leant forward and switched on the radio receiver again. 'I don't want you to be able to back out of this either, so, in front of witnesses, Brontë O'Brian, will you marry me?'

'Yes, Elijah Munroe, I will,' she said, and saw him switch off the receiver again. 'So, it's official now, neither of us can back out.'

'Nope. We're stuck with each other,' he replied, and then the corners of his mouth tipped up ever so slightly. 'So, what now?'

'What do you mean, what now?' she asked, and saw his smile widen.

'Well, we're stuck in this ambulance with nothing to do until the breakdown truck arrives.'

'We can listen to the radio,' she suggested. 'Earwig on other paramedics.'

'Nope, that's not lighting my fire.'

'Or we could play I Spy,' she observed. 'I can think of some pretty fiendish ones.'

'I can think of better games to play,' he said, sliding his arm along her seat, and she shook her head.

'I'm sure you can, but we're not playing any of those, not in an ambulance.'

He stuck out his tongue at her. 'Spoilsport.'

'Realist, more like.' She grinned. 'I'm already notorious on the radio, and I've no intention of becoming even more so should the breakdown truck arrive.'

'We could just abandon the ambulance, go back to your place?'

She shook her head at him.

'Eli Munroe, you're a very dangerous man.'

He laughed. 'Nah. Big pussycat, me. So what do you say—will we go back to your place? I mean, we could be sitting here all night, getting colder, and colder, and you have a lovely, warm and very cosy bed....'

It sounded so tempting, and his eyes were hot, and as he'd said, they had no idea when the breakdown truck would arrive. In fact, when she'd told Dispatch of the problem with the ambulance, the caller had been anything but encouraging.

'Okay, we'll get a taxi back to my place,' she said, but as she made to open her door, Eli stayed her arm.

'Just one question before we go,' he declared. 'Do you have a tape measure at home?'

'I...I think so,' she said, bewildered. 'In fact, I'm almost positive I do. Why?'

'Because there's one question you still haven't given me the answer to.' He grinned. 'So, in about half an hour's time, when I've got you naked in front of me...'

'Are you naked in this scene, too?' she said, her lips curving.

'Oh, absolutely.' He nodded. 'Most definitely.'

'So, when I'm naked, and you're naked?' she prompted.

'With your tape measure, I'll finally get to find out what your hip measurement is.'

'I promise you the results will be ugly.' She smiled. 'In fact, it might make you reconsider your offer to marry me.'

He lifted her hand to his lips, his blue eyes soft, and warm.

'Even if you turn out to have hips the size of a barn door—'

'*Hey!*'

'Nothing is going to stop me from marrying you, Brontë O'Brian. We're a partnership now in every sense of the word, and we always will be.'

And as she gazed back at him, and saw the love in his eyes, she knew they would be.